Topics in Operational Research

A series of books under the advisory editorship of Professor S. Vajda of the University of Birmingham

Planning by Mathematics

Topics in Operational Research

Since the end of the Second World War, studies in the field of Management Sciences and Operational Research have developed at an extraordinary speed, and have gained in width and depth.
The present series of books on "Topics in Operational Research" contains contributions from authors who have special expertise in this field. There exist, of course, excellent texts on the theoretical background, but this series fulfils a need for detailed descriptions of simple numerical or algebraic techniques, and for reports on practical experience with these techniques, useful to managers and executives in many branches of social, commercial, and industrial activity. Knowledge of such techniques will be spread more widely by the publication of these monographs, and further development stimulated thereby.
The texts are written in such a way that little, if any, advanced knowledge from any branch of science will be required of the reader and no mathematical knowledge beyond A-level G.C.E. or O.N.C.
The various books in the series should thus be of interest not only to management and engineering personnel, but also to teachers in the later forms in schools, and in colleges, as suggestions for possible syllabuses for introductory courses and as sources of interesting problems and examples.

S. VAJDA

Planning by Mathematics

S. VAJDA

Dr. Phil.

Department of Transportation and Environmental Planning,
The University of Birmingham

Second Edition

Pitman Publishing

First published 1969
Second edition 1973

SIR ISAAC PITMAN AND SONS LTD.
Pitman House, Parker Street, Kingsway, London, WC2B 5PB
P.O. Box 46038, Portal Street, Nairobi, Kenya

SIR ISAAC PITMAN (AUST.) PTY. LTD.
Pitman House, 158 Bouverie Street, Carlton, Victoria 3053, Australia

PITMAN PUBLISHING COMPANY S.A. LTD.
P.O. Box 11231, Johannesburg, S. Africa

PITMAN PUBLISHING CORPORATION
6 East 43rd Street, New York, N.Y. 10017, U.S.A.

SIR ISAAC PITMAN (CANADA) LTD.
495 Wellington St West, Toronto 135, Canada

THE COPP CLARK PUBLISHING COMPANY
517 Wellington St West, Toronto 135, Canada

Cased edition ISBN 0 273 31767 9
Paperback edition ISBN 0 273 31766 0

Text set in 10/11pt. Monotype Times, printed by letterpress,
and bound in Great Britain at The Pitman Press, Bath

G2—(T.1263/1368: 75)

Preface

This book contains a completely revised version of *Readings in Mathematical Programming*, which itself was the second edition of *Readings in Linear Programming*.

The previous editions were, to use a descriptive term, technique-oriented, i.e. each chapter dealt with a special mathematical technique of Operational Research, and these techniques were illustrated by appropriate examples of their applications. The present edition is problem-oriented. The twenty-one chapters deal with applied, realistic problems, though of course in a simplified form. For each problem we draw on our knowledge of one or more techniques, and examine their suitability for solving the problem on hand. Many more techniques are referred to than in earlier editions, which reflects the still rapidly expanding scope of the subject.

The two approaches are not necessarily contradictory, and a few chapters have been taken over from the previous edition, with little or no alteration.

As regards the techniques applied, it is assumed that the Simplex Method is fairly well known. Nevertheless, it will be found convenient to refer occasionally to the Appendix, which describes briefly and rather sketchily the Simplex and Dual Simplex Methods, the Inverse Matrix Method, the Decomposition routine and Integer routines of linear programming, as well as two methods of quadratic programming. For background, details and proofs the reader should consult the literature, particularly *Mathematical Programming in Practice* by E. M. L. Beale, in this Topics in Operational Research series.

Other techniques, belonging to the wider field of mathematical programming, including Dynamic Programming (for example fractional programming, or branch and bound methods), are used and explained in the text where required. References are made in all chapters to the selective bibliography which is appended.

The author believes, on the basis of some extensive experience in teaching the subject, that each chapter that follows may well serve as a topic for one—shorter or longer—instructive talk on the use of Operational Research Techniques in industry, commerce or some other economic activities.

Birmingham, 1968 S. VAJDA

Contents

1 Transportation With and Without Transhipment

In a small town a need arises for a number of relief buses. In particular, 3, 3, 4, and 5 buses are required at points A, B, C, and D respectively. These 15 buses come from the garages G_1, G_2, and G_3 where 2, 6, and 7 buses are ready for such an emergency. The manager would like to distribute the buses in such a way that the total of bus-minutes from the garages to their destinations is as small as possible. The time in minutes taken to travel from each garage to each destination is given in the following table:

	To	A	B	C	D
From G_1		13	11	15	20
G_2		17	14	12	13
G_3		18	18	15	12

If we denote the number of buses to be sent from G_i to A, B, C, and D respectively by x_{i1}, x_{i2}, x_{i3}, and x_{i4}, then the requirements can be expressed as follows:

(*a*) The buses at the garages are distributed among the destinations:

$$x_{11} + x_{12} + x_{13} + x_{14} = 2$$
$$x_{21} + x_{22} + x_{23} + x_{24} = 6$$
$$x_{31} + x_{32} + x_{33} + x_{34} = 7$$

(*b*) The buses come to the destinations from various garages:

$$x_{11} + x_{21} + x_{31} = 3$$
$$x_{12} + x_{22} + x_{32} = 3$$
$$x_{13} + x_{23} + x_{33} = 4$$
$$x_{14} + x_{24} + x_{34} = 5$$

The "cost" to be minimized is the total time taken measured in bus-minutes, i.e.

$$13x_{11} + 11x_{12} + 15x_{13} + 20x_{14} + \ldots + 12x_{34}$$

This is a very simple case of linear programming. All the coefficients are unity or zero, and the pattern of variables occurring in any single equation is a very special one. Of the 7 equations only 6 are linearly independent, and thus a basic feasible solution will consist of variables with not more than 6 positive values.

It is easy to find such a basic feasible solution. Consider the following table which will be filled in:

		Numbers at destinations			
		3	3	4	5
Numbers	2				
at	6				
sources	7				

To begin with, take the cell of shortest time, which is that of x_{12} with its row and column labelled 2 and 3. Because 2 is the smaller, enter 2 into this cell, thereby allocating the 2 buses from G_1. We need one more bus at B, and thus we are now concerned with the reduced scheme:

		A	B	C	D
		3	1	4	5
G_2	6				
G_3	7				

Dealing with this table as we did with the previous one, we enter the smaller of 4 and 6 into the cell of x_{23}. Proceeding in the same way, the entries in the table are eventually:

	3	3	4	5
2	.	2	.	.
6	1	1	4	.
7	2	.	.	5

The totals of rows and columns are, of course, those originally given. The total time of this scheme is 197 bus-minutes, though we do not yet know whether this is the best possible scheme, i.e. that producing the shortest total travelling time.

In order to investigate this point, consider any empty cell, e.g. that of x_{11}, and find out whether a basic feasible solution, with an entry in that cell, would give a better result. If we entered there 1, say, then we must take 1 away from x_{12}; to balance again, we must add 1 to x_{22} and then subtract 1 from x_{21}. Performing these transfers, we change the total cost by

$$c_{11} - c_{12} + c_{22} - c_{21} = 13 - 11 + 14 - 17 = -1$$

i.e. the total cost would be reduced. Here c_{ij} is the time taken to travel from G_i to the *j*-th destination. On the other hand, if we wanted a positive entry for x_{14}, the cost would be changed by

$$c_{14} - c_{34} + c_{31} - c_{21} + c_{22} - c_{12} = 12$$

so that this would produce an increase.

In this way we could examine each empty cell and find out whether the scheme could be improved. This method was called the "stepping-stone method" by Charnes and Cooper [7]. However, it would be very awkward if we had to make such an investigation for each empty cell, finding every time a circuit that leads from the empty cell by horizontal and vertical steps using only filled-in cells (because we consider only *basic* feasible solutions), and that returns to the empty cell considered. A short-cut is provided by the "modified" method, which works as follows: on the margins of the table we enter fictitious costs (also called "shadow" costs) c_i for every row and d_j for every column, such that $c_i + d_j = c_{ij}$ for all occupied cells. These fictitious costs can be constructed by making c_1, say, equal to 0 whereby the others are implicitly fixed. (These fictitious costs need not be non-negative.) Consider now the first transfer mentioned above. A transfer of one bus changes the total cost by

$$c_{11} - c_1 - d_2 + c_2 + d_2 - c_2 - d_1 = c_{11} - c_1 - d_1$$

Hence, if the cost of the unoccupied cell implied by the fictitious cost $c_1 + d_1$ is larger than the true cost c_{11}, then such a rearrangement leads to an improvement. The fictitious costs indicate, in the same way, whether an improvement is indicated, even if the transfers involve more than four cells, as in our second example. This is so because each more involved transfer can be made up of a succession of simple ones.

If the implied fictitious cost equals the true cost, then an entry could be made in that cell, but the resulting reallocation would leave the total cost unchanged.

For the procedure which we have outlined to work, the number of occupied cells must equal the number of independent equations in the transportation problem, i.e. $n + m - 1$, where n and m are the numbers of sources and destinations respectively.

In our problem the fictitious costs (f.c.) of rows and columns, and the implied fictitious costs of the cells unoccupied in our first solution, are given by the following table:

					f.c.
	14	.	9	8	0
	.	.	.	11	3
	.	15	13	.	4
f.c.	14	11	9	8	

Comparison with the true costs shows that it is worth while making a transfer into x_{11} (true cost 13, fictitious cost 14) and to effect this we alternately add to and subtract from x_{11}, x_{21}, x_{22}, x_{12}. We cannot transfer more than 1, because otherwise x_{21} would become negative. We obtain thus the following new solution, with the new fictitious costs as indicated:

	3	3	4	5	f.c.
2	1	1	.	.	0
6	.	2	4	.	3
7	2	.	.	5	5
f.c.	13	11	9	7	

If we compare the implied fictitious costs in the empty cells with their true costs, we find that the former are everywhere the smaller, and hence we have obtained the final, cheapest, solution. The total travelling time involved is 196 bus-minutes.

If there is more material at all sources taken together than there is required at all the destinations, e.g. when there is a supply of buses at the garages which exceeds the number of those actually needed at A, B, C, and D, then we add another, fictitious, destination, to which all surplus of material is deemed to be directed, at zero cost. The buses earmarked for this dummy destination would, of course, actually not be moved at all from the garages.

The algebraic formulation of the Transportation Problem was as follows:

Maximize $\sum_i\sum_j c_{ij}x_{ij}$ with non-negative x_{ij}, subject to

$$\sum_i x_{ij} = b_j \qquad \text{and} \qquad \sum_j x_{ij} = a_i$$

If the constraints were replaced by

$$\sum_i x_{ij} \leqslant b_j \qquad \text{and} \qquad \sum_j x_{ij} \leqslant a_i$$

then we could re-establish the simplicity of the transportation model by writing for the inequalities

$$\sum_i x_{ij} + y_j = b_j \qquad \sum_j x_{ij} + x_i = a_i$$

and adding the following two equations, which do not restrict the variables any further

$$\sum x_i + z = \sum a_i \qquad \sum y_j + z = \sum b_j$$

We have thereby established a transportation scheme, defined by the table

						totals
	x_{11}	x_{12}		x_{1m}	x_1	a_1
	—	—	—	—	—	—
	x_{n1}	x_{n2}		x_{nm}	x_n	a_n
	y_1	y_2		y_m	z	$\sum b_j$
totals	b_1	b_2		b_m	$\sum_i a_i$	

The variable z is simply the total of all x_{ij}. Note also that the sum of all entries in the marginal column equals that of the entries in the marginal row, even if the sum of all a_i is not equal to that of all b_j.

Transhipment

We now consider the possibility that the times given in the table for the distribution of buses to their destinations were shortest routes only subject to the condition that no bus should pass another garage, or the destination of another bus (to avoid traffic jams), even if the length of the route could thereby be shortened.

Let us now ignore this restriction and allow any bus to call at intermediate stops. (A similar situation clearly arises when shipping is allowed to call at intermediate ports, whence the name of the problem.) We are then also interested in the times taken between garages, or between any of the ultimate destinations. Therefore we extend the data as follows:

Times between two points

	G_1	G_2	G_3	A	B	C	D
G_1	0	6	5	13	11	15	20
G_2	5	0	4	17	14	12	13
G_3	4	4	0	18	18	15	12
A	14	19	20	0	5	4	5
B	10	12	17	5	0	2	8
C	17	13	14	4	2	0	9
D	18	12	14	5	8	9	0

It will be noticed that this table is not symmetrical. For instance, we assume that to go from A to G_3 takes 20 minutes (on the direct route), but from G_3 to A only 18 minutes. Such differences can easily be explained by the exigencies of the traffic, such as one-way streets, etc. A scrutiny of the table will also show that it is shorter to go from G_1 to D via G_3 than directly, and similar features attach to other connections. This is realistic, in particular if direct transport without intermediate calls is not possible.

We must now fix the totals of sources and destinations, and here a difficulty arises. It would not be useful to try and solve the problem for the same totals as before because then the requirements at the garages would be zero, and the totals at the points A, B, C, and D, taken as sources, would also be zero. Thus no "transhipment" would ever arise.

To overcome this difficulty we add to all sources and to all destinations of the new table a number large enough to cover all potential transhipments. For instance, we could choose for this the total of all buses in the scheme. Then we solve the new transportation problem. In the present example, we add 20 to all original totals and obtain the following pattern:

	20	20	20	23	23	24	25	f.c.
22	20	.	.	1	1	.	.	0
26	.	20	.	.	2	4	.	3
27	.	.	20	2	.	.	5	5
20	.	.	.	20	.	.	.	−13
20	.	.	.	.	20	.	.	−11
20	.	.	.	.	.	20	.	−9
20	.	.	.	.	.	.	20	−7
f.c.	0	−3	−5	13	11	9	7	

We have also entered, as a first feasible solution, that solution which was optimal in the earlier case, and added 20 to all diagonal cells. The total cost is, of course, the same as in the earlier case, because the entries in the diagonal do not add anything to the cost.

Testing for optimality, by the use of fictitious costs which are also shown, we find that a transfer into the cell (G_3, G_1) would reduce the cost, and we obtain then:

	20	20	20	23	23	24	25	f.c.
22	18	.	.	3	1	.	.	0
26	.	20	.	.	2	4	.	3
27	2	.	20	.	.	.	5	4
20	.	.	.	20	.	.	.	−13
20	.	.	.	.	20	.	.	−11
20	.	.	.	.	.	20	.	−9
20	.	.	.	.	.	.	20	−8
f.c.	0	−3	−4	13	11	9	8	

This is the optimal solution, and the answer to the transhipment problem is now found by omitting the entries in the diagonal. A comparison with the previous result is interesting:

	Without transhipment	*With transhipment*
G_1 to A	1	3 ←
G_1 to B	1	1
G_2 to B	2	2
G_2 to C	4	4
G_3 to A	2	—
G_3 to D	5	5
G_3 to G_1	—	2
Total time in bus-minutes	196	194

The transportation problem was first mentioned in [25], and the transhipment was solved in [32].

2 Bid Evaluation

A Government agency requires 300 shirts at depot A, 400 at depot B, and 300 at depot C. Manufacturer X is prepared to supply up to 800, manufacturer Y up to 500, and manufacturer Z up to 300. The cost of the supplies depends on the prices at the factory and on the cost, to the Government, of transporting the goods to the depots. Quoting prices, manufacturer X makes them dependent on the total bought from him. All relevant costs are given in the following table.

Cost (in some convenient unit) to the Government of 100 shirts delivered at the following depots:

From		A	B	C
X	up to 200	9	11	13
	up to another 600	10	12	14
Y	up to 500	12	15	15
Z	up to 300	13	14	16

It is required to make awards to the suppliers in such a way that the cost to the Government is as low as possible.

In order to make use of the simple transportation routine, we imagine a further depot, D (for dummy), to which goods offered but not required are sent, at zero cost. We can cope with the price difference in X's offer by imagining two manufacturers, X_1 and X_2, with prices as given in the two rows. This is satisfactory, because on minimizing the total cost we shall automatically only make use of the more expensive goods after the exhaustion of the total available at the cheaper prices.

Solving by the methods available for the transportation problem, we obtain (in units of 100 shirts):

		3	4	3	6	f.c.
X_1	2	2	.	.	.	0
X_2	6	1	4	1	.	1
Y	5	.	.	2	3	2
Z	3	.	.	.	3	2
	f.c.	9	11	13	−2	

The total cost is 120 times 100.

We now turn to the consideration of complications of the types that might occur in practice.

Manufacturer Y is not prepared to deliver less than 400 and, as already stated, not more than 500.

We note that the solution above allocates to him only 200. We cannot cope with his stipulation by introducing a constraint demanding that his total delivery to all genuine depots, i.e. excluding D, must be at least 400, because this would exclude the possibility of rejecting his offer altogether. His condition means that his delivery is between 400 and 500 (both limits included) or nothing. If the solution above had included an amount within these conditions, then clearly this would be the cheapest under the circumstances. But as it is, we must compute the cost of buying goods only from the other manufacturers, and the cost when Y is being allocated precisely 400. The smaller of these costs will then decide the issue. It can be proved analytically that any delivery from Y of another amount is bound to be more expensive in total cost.

Ignoring Y altogether, we obtain the following optimal feasible solution:

		A	B	C	D	f.c.
		3	4	3	1	
X_1	2	2	.	.	.	0
X_2	6	1	4	1	.	1
Z	3	.	.	2	1	3
	f.c.	9	11	13	−3	

The cost is 122.

To compute the lowest cost under the assumption that Y delivers precisely 400, we assign a very high price to his deliveries to D, so that nothing will go from him into that dummy depot. We have then to solve the problem given by the cost matrix (with requirements on the margins).

Cost Matrix

		A	B	C	D
		3	4	3	5
X_1	2	9	11	13	0
X_2	6	10	12	14	0
Y	4	12	15	15	M
Z	3	13	14	16	0

M is a number which is taken to be larger than any other with which it is to be compared during the computation.

The solution is

		3	4	3	5	f.c.
X_1	2	2	.	.	.	−1
X_2	6	.	4	.	2	0
Y	4	1	.	3	.	2
Z	3	.	.	.	3	0
	f.c.	10	12	13	0	

The total cost of this scheme is 123. Since this is higher than the cost when Y is altogether excluded, the latter scheme, with a total cost of 122, will be accepted.

As a matter of interest, we also work out the cost of allowing Y to deliver precisely 500. This will, of course, be higher than 123. The cost table to be used is now

		3	4	3	6
X_1	2	9	11	13	0
X_2	6	10	12	14	0
Y	5	12	15	15	M
Z	3	13	14	16	0

and an optional scheme is

		3	4	3	6	f.c.
X_1	2	1	1	.	.	0
X_2	6	.	3	.	3	1
Y	5	2	.	3	.	3
Z	3	.	.	.	3	1
	f.c.	9	11	12	−1	

with a total cost of 125.

To complicate matters further, we assume now that Z is prepared to supply more than 300, and that he even reduces his price for larger deliveries as follows:

	A	B	C
Up to 300 . . .	13	14	16
Excess over 300, up to 600 . .	11	12	14
Excess over 600, up to 900 . .	10	11	13

Once more it is impossible to arrive at the answer in one single process. The difficulty arises from the fact that we cannot make certain that the cost for the higher amounts comes into operation only after the lower amounts have been used up at the higher prices—as we could do for the manufacturer who demanded higher prices for larger amounts. Our procedure must be as follows: we obtain the cost under the assumption that only an amount involving the highest prices is accepted. We have just worked out the answer to this problem; it was 122, and to obtain it we had to compare two possibilities for the allocation to Y.

Then, we determine the cheapest cost of an allocation of between (more than) 300 and 600 and, finally, of an allocation of between (more than) 600 and 900. Accordingly, our next step is concerned with making certain that 300 at any rate are allocated to Z. This is achieved by introducing again a very high cost of delivery to D. We do not worry at this stage about Y's contribution, since it might sort itself out anyway. The problem below must thus be solved:

Cost Matrix

		A	B	C	D
		3	4	3	9
X_1	2	9	11	13	0
X_2	6	10	12	14	0
Y	5	12	15	15	0
Z_1	3	13	14	16	M
Z_2	3	11	12	14	0

It is easily verified that an optimal solution is

		3	4	3	9	f.c.
X_1	2	2	.	.	.	0
X_2	6	1	4	.	1	1
Y	5	.	.	.	5	1
Z_1	3	.	.	3	.	3
Z_2	3	.	.	.	3	1
	f.c.	9	11	13	−1	

Y's condition need not concern us, because his offer is entirely rejected in any case. The total cost of this scheme would be 124, so that we need not consider it any more.

Finally we solve the following problem:

Cost Matrix

		A 3	B 4	C 3	D 12
X_1	2	9	11	13	0
X_2	6	10	12	14	0
Y	5	12	15	15	0
Z_1	3	13	14	16	*M*
Z_2	3	11	12	14	*M*
Z_3	3	10	11	13	0

by the allocation

		A 3	B 4	C 3	D 12	f.c.
X_1	2	2	.	.	.	0
X_2	6	1	.	.	5	1
Y	5	.	.	.	5	1
Z_1	3	.	3	.	.	4
Z_2	3	.	1	2	.	2
Z_3	3	.	.	1	2	1
	f.c.	9	10	12	−1	

As before Y is excluded from the contracts. The total cost is 123, which is still too high. Our conclusion is, therefore, that the scheme excluding Y, and not taking more than 300 from Z, is cheapest, with a cost of 122.

In these examples it has been implicitly assumed that the differences between the costs, to the Government, of unit deliveries to the various depots from the same manufacturer are due to the transport costs, which are borne by the Government; the costs at the factories are the same, wherever the goods go. Thus, when a manufacturer offers different prices for different quantities, the differences between these quantities are the same at all (true) depots. If this assumption holds, then the computations can be slightly simplified, by avoiding large tables. We show this for the last of the alternatives for Z. It is, in fact, sufficient to assume that Z charges for all units the same, lowest, price:

		A 3	B 4	C 3	D 12
X_1	2	9	11	13	0
X_2	6	10	12	14	0
Y	5	12	15	15	0
Z	9	10	11	13	0

This is solved by

		3	4	3	12	f.c.
X_1	2	2	.	.	.	0
X_2	6	1	.	.	5	1
Y	5	.	.	.	5	1
Z	9	.	4	3	2	1
	f.c.	9	10	12	−1	

The apparent cost is now 111, but to this must be added 3 times 1, and 3 times 3, resulting in 123, as before.

If the columns marked A, B, etc., do not stand for different depots, but for different goods, e.g. shirts, socks, etc., then we have no reason to expect that differences between rows relating to the same manufacturer will be constant. In such a case further complications can arise through, say, ceilings on the total deliveries of shirts, and of socks, but not of both together. Furthermore, the transportation model breaks down and a general routine (e.g. the Simplex Method) must be applied.

The problem of evaluation of tenders to the U.S. Government is treated in [33].

3 Investment

The Financial Manager of a firm wants to invest an amount of 100 (in some appropriate units, say £1,000) so as to maximize its yield. He considers the following possibilities:

Type	A_1	A_2	B_1	B_2	C_1	C_2
Yield p.a.	3%	2·5%	3·5%	4%	5%	4·5%
Invested amount	x_1	x_2	y_1	y_2	z_1	z_2

He is not entirely free in his choice, because it is the firm's policy that at least 40 per cent of the whole amount must be invested in units of type A, and not more than 35 per cent in any single one of the other two types.

The correct procedure is rather obvious from first principles, but the example serves to illustrate it for more complex cases. The conditions for investment are as follows:

$$x_1 + x_2 + y_1 + y_2 + z_1 + z_2 = 100 \qquad x_1 + x_2 - t_1 = 40$$
$$y_1 + y_2 + t_2 = 35 \qquad z_1 + z_2 + t_3 = 35$$

Maximize $3x_1 + 2{\cdot}5x_2 + 3{\cdot}5y_1 + 4y_2 + 5z_1 + 4{\cdot}5z_2 = B$, say.

The t_i were introduced in order to convert the inequalities into equations.

It is of interest to note that this problem can be transformed into one of transportation type. For this purpose, we split x_1 and x_2, which add up to at least 40, into two portions,

$$x_1 = x_{10} + x_{11} \qquad \text{and} \qquad x_2 = x_{20} + x_{21}$$

so that

(1) $x_{10} + x_{20} = 40$, and hence
(2a) $x_{11} + x_{21} + y_1 + y_2 + z_1 + z_2 = 60$

We introduce, then, $y_0 = y_1 + y_2$ and $z_0 = z_1 + z_2$, so that (2a) becomes

(2) $x_{11} + x_{21} + y_0 + z_0 = 60$

Moreover, since

(3) $y_1 + y_2 + t_2 = 35$ and (4) $z_1 + z_2 + t_3 = 35$

we have

(5) $y_0 + t_2 = 35$ and (6) $z_0 + t_3 = 35$

We introduce slack variables u, v, and w by writing

(7) $x_{10} + x_{11} + u_1 = 100$ (8) $x_{20} + x_{21} + u_2 = 100$
(9) $y_1 + v_1 = 35$ (10) $y_2 + v_2 = 35$
(11) $z_1 + w_1 = 35$ (12) $z_2 + w_2 = 35$

If we now change the signs in the equations (5) to (12) and add all 12 equations, then we obtain

(13) $-u_1 - u_2 - v_1 - v_2 - w_1 - w_2 = -240$

In the 13 equations every variable appears twice, once with a positive and once with a negative unit coefficient. Such a case can be written in the form of a transportation problem. In fact, the present case is relatively simple, because every equation contains either only positive or only negative terms. We can write it in the following form, where the variables mentioned add up to the marginal horizontal and vertical totals:*

	100	100	35	35	35	35	35	35
40	x_{10}	x_{20}						
60	x_{11}	x_{21}	y_0	z_0				
35			t_2		y_1	y_2		
35				t_3			z_1	z_2
240	u_1	u_2			v_1	v_2	w_1	w_2

The yields (negative costs) are as follows (in per cent):

3	2·5						
3	2·5	0	0				
		0		3·5	4		
			0			5	4·5
0	0			0	0	0	0

In the cells where no variable is entered the yield is prohibitively low. We must remember that we want to maximize the total yield, not to minimize a total cost, as is usual in the transportation case.

* This transformation is based on a suggestion by E. M. L. Beale.

We start with the following table, constructed according to the usual rules:

	100	100	35	35	35	35	35	35
40	+		10			30^-		
60			25	35				
35					30^-	5^+		
35							35	0
240	100^-	100			5^+			35

For the construction of this table, we started with the cell of the highest yield and entered 35. Thus we completed, at the same time, a row and a column. In order to obtain a basic solution, which needs, in this case, 12 basic variables, i.e. 12 entries, we entered a 0 in the cell of the covered row, with the next highest yield.*

It is obvious, even without quoting the shadow costs, that the entry 30 in the first row, with an unacceptably large negative yield, must be removed, and we do this by using the empty cell of that row with the highest yield. The $^+$ and $^-$ marks indicate the ensuing changes. The next table is:

	100	100	35	35	35	35	35	35
40	30^+		10^-					
60			25	35				
35			+		0^-	35		
35							35	0
240	70^-	100			35^+			35

There were two cells with 30^- in the previous tableau. Had we removed both these entries, then we should have been left with only 11 entries in the table. To avoid this, we remove only the entry in the first row, and change the other entry to 0.**

Clearly, the entry 10 in the first row should be removed. The $^+$ and $^-$ marks show again the necessary transfers. However, they indicate a transfer of 0 only, so that the entries remain the same, except for a 0 in the third row and third column, and no entry in the third row and fifth column.

* Eisemann [14] has shown that, if a row and a column are simultaneously covered, the row should be considered as having a residue of 0 left.
** cf. Eisemann [14].

The next step, still concerned with removing 10 from its place in the first row, leads to

	100	100	35	35	35	35	35	35	f.c.
40	40								0
60			25	35					1
35			10			25			1
35							35	0	1·5
240	60	100			35	10		35	−3
f.c.	3	3	−1	−1	3	3	3·5	3	

The shadow costs indicate that we have now reached the final answer. (Remember, again, that we want to *maximize* the total yield.) It means that the Financial Manager should invest

40 in A_1
25 in B_2
35 in C_1

The fact that $t_2 = 10$ means that type B of investment falls 10 short of the ceiling imposed on it. The total yield will be 3·95%.

4 Routing Aircraft

The following numbers of passengers should be transported by a fleet of aircraft, on three different routes, within one year:

Route	Numbers
1	320,000
2	165,000
3	190,000

There are three types of aircraft available: 15 of type I, 14 of type II, and 18 of type III. The annual cost of the operation of one of them is given in the following:

Cost Table

	Aircraft type		
On route	I	II	III
1	120,000	130,000	180,000
2	120,000	130,000	150,000
3	160,000	110,000	140,000

The capacity of the aircraft to carry passengers depends on the route and is given in the following:

Capacity Table (passengers per year)

	Aircraft type		
On route	I	II	III
1	20,000	15,000	14,000
2	18,000	13,000	10,000
3	12,000	14,000	8,000

If a passenger turns up and can not be accommodated, then a loss is incurred which has been determined by the accountants to amount to

150 per passenger on route 1
150 per passenger on route 2
80 per passenger on route 3.

It is required to allocate aircraft to routes in such a way that the total cost is minimized.

Denote the number of aircraft of type j on route i by x_{ij}, and the number of passengers who cannot be accommodated on route i by $1{,}000x_i$. Then we have the constraints (after division by 1,000),

$$\begin{aligned} 20x_{11} + 15x_{12} + 14x_{13} + x_1 &= 320 \\ 18x_{21} + 13x_{22} + 10x_{23} + x_2 &= 165 \\ 12x_{31} + 14x_{32} + 8x_{33} + x_3 &= 190 \end{aligned}$$

and, because of the numbers of aircraft available,

$$\begin{aligned} x_{11} + x_{21} + x_{31} &= 15 \\ x_{12} + x_{22} + x_{32} &= 14 \\ x_{13} + x_{23} + x_{33} &= 18 \end{aligned}$$

Note that the number of aircraft on a particular route may be fractional. This would mean that some aircraft would fly at larger intervals than was assumed when the carrying capacity was established.

It is required to minimize (after division by 10,000)

$$\begin{aligned} &12x_{11} + 13x_{12} + 18x_{13} + 12x_{21} + 13x_{22} + 15x_{23} + 16x_{31} \\ &+ 11x_{32} + 14x_{33} + 15x_1 + 15x_2 + 8x_3 \end{aligned}$$

The scheme is similar to that of the transportation problem, but the coefficients in the first group of equations are not all 0 or 1 in the present case. Even so, the procedure for solving the problem has certain similarities with that for the transportation problem in its classical form.

We exhibit the problem in compact form by entering, in the following table, the coefficients of the first group of constraints, p_{ij} say ("capacities"), in the lower left-hand corner, and those of the objective function, c_{ij} say ("costs"), in the upper right-hand corner:

		Required on routes		
		1 320	2 165	3 190
Available of types	I 15	12 20	12 18	16 12
	II 14	13 15	13 13	11 14
	III 18	18 14	15 10	14 8
	Loss	15 1	15 1	8 1

Note that the last row has no restriction on the total of its entries.

To find a basic feasible solution, we proceed as follows: consider the top cell in the first column. We enter there either 15 (number of aircraft available), or 320/20 (number of aircraft required to carry all the passengers on that route), whichever is smaller. Thus, we enter $x_{11} = 15$.

This exhausts the first row, but leaves a further requirement in the first column, amounting to $320 - 15 \times 20 = 20$. Therefore, we enter in the next cell of the first column, the smaller of 14 and 20/15, viz. $x_{12} = 4/3$. Now the whole of the first column is satisfied, but not the second row. Continuing in this fashion, we enter

$$x_{22} = \min [14 - 4/3, 165/13] = 38/3$$
$$x_{23} = \min [18, (165 - 38/3 \times 13)/10] = 1/30$$
$$x_{33} = \min [18 - 1/30, 190/8] = 539/30$$
$$x_{3} = \min [190 - 539/30 \times 8] = 694/15$$

Thus

	320	165	190
15	15		
14	4/3	38/3	
18		1/30	539/30
Loss			694/15

This is our first basic feasible solution. The total cost is $984\frac{1}{6}$.

Now we must find out if this is the best possible schedule, and if it is not, how to improve it. We use again shadow costs, denoting that of row j by r_j, and that of column i by s_i. However, this time they are affected by the capacities p_{ij}, which appear as coefficients in the first set of constraints.

Let us take such shadow costs that

$$s_i p_{ij} + r_j = c_{ij}$$

for those cells (i, j) for which the entry x_{ij} is positive. Imagine, without restriction of generality, that $x_{11} = 0$ while x_{12}, x_{21} and x_{22} are positive. We investigate by which amount (positive or negative) the total cost changes if we put θ into the cell of x_{11}. This leads to changes in other cells, because the constraints must remain satisfied.

In order to retain the row totals, x_{21} must be changed into $x_{21} - \theta$. To satisfy the column total, we change x_{12} into $x_{12} - \theta p_{11}/p_{12}$, so that we have

$$x_{11}p_{11} + x_{12}p_{12} = (x_{11} + \theta)p_{11} + (x_{12} - \theta p_{11}/p_{12})p_{12}$$

To satisfy the other row total, we change x_{22} into $x_{22} + \theta p_{11}/p_{12}$. Now the second column total must be reconstructed, and we obtain this by adjusting x_2, the slack variable in the corresponding constraint. This does not involve any further adjustment, because the total of the row of slacks is not prescribed.

We change x_2 into $x_2 + \theta p_{21} - \theta p_{11}p_{22}/p_{12}$, because thereby

$$\begin{aligned} x_{21}p_{21} + x_{22}p_{22} + x_2 = (x_{21} - \theta)p_{21} &+ (x_{22} + \theta p_{11}/p_{12})p_{22} \\ &+ (x_2 + \theta p_{21} - \theta p_{11}p_{22}/p_{12}) \end{aligned}$$

To see if such addition to $x_{11} = 0$ and the consequential adjustments in the other four cells are useful, we compute the amount thereby added to the objective function (the total cost). This amount equals

$$\begin{aligned} &\theta[c_{11} - c_{21} - c_{12}p_{11}/p_{12} + c_{22}p_{11}/p_{12} \\ &\qquad\qquad + c_2(p_{21} - p_{11}p_{22}/p_{12})] \\ &= \theta[c_{11} - (s_2p_{21} + r_1) - (s_1p_{12} + r_2)p_{11}/p_{12} \\ &\qquad\qquad + (s_2p_{22} + r_2)p_{11}/p_{12} + c_2(p_{21} - p_{11}p_{22}/p_{12})] \\ &= \theta[c_{11} - (s_1p_{11} + r_1) + (s_2 - c_2)(p_{11}p_{22}/p_{12} - p_{21})] \end{aligned}$$

Hence, if we decide to make the shadow cost of the row of the slacks zero, so that the cost of a cell in that row equals the shadow price of the column (in the present example $s_2 = c_2$), then it is useful to add a positive amount θ to $x_{11} = 0$, if and only if

$$s_1p_{11} + r_1 > c_{11}$$

In general, we enter θ into an empty cell (i, j), if $s_i p_{ij} + r_j > c_{ij}$. The demonstration refers, strictly speaking, only to a rectangle such as was formed by the cells (1,1), (1,2), (2,1) and (2,2), but can easily be extended to adjustments in more cells, provided that after having chosen an empty cell which receives a useful entry, the other cells to be adjusted have already positive amounts in them.

In the present case the determination of the shadow costs is somewhat more complicated than it was in the classical case of Chapter 1, where no p_{ij} were involved. We have now to solve the following system of linear equations:

$$20s_1 + r_1 = 12 \qquad 15s_1 + r_2 = 13 \qquad 13s_2 + r_2 = 13$$
$$10s_2 + r_3 = 15 \qquad 8s_3 + r_3 = 14 \qquad s_3 = 8$$

They are solved by

s_1	s_2	s_3	r_1	r_2	r_3
169/30	$6\frac{1}{2}$	8	$-100\frac{2}{3}$	$-71\frac{1}{2}$	-50

For the empty cells above, we compute $s_i p_{ij} + r_j$, and the differences between these values and the true costs:

$s_i p_{ij} + r_j$

	$i = 1$	2	3
$j = 1$		$16\frac{1}{3}$	$-4\frac{2}{3}$
2			$40\frac{1}{2}$
3	866/30		
	169/30	$6\frac{1}{2}$	

Differences

	$4\frac{1}{3}$	$20\frac{2}{3}$
		$29\frac{1}{2}$
326/30		
—281/30		

We can expect to gain most by making x_{32} positive. Enter $x_{32} = \theta$. This will reduce the total cost by $29{\cdot}5\theta$. As a consequence, we have

$$x_{22} = 38/3 - \theta \qquad x_{32} = \theta$$
$$x_{23} = 1/30 + 13\theta/10 \qquad x_{33} = 539/30 - 13\theta/10$$
$$x_2 = 694/15 - 18\theta/5$$

To keep the solution feasible and basic, we make one of these five entries zero, while the others remain non-negative. Therefore θ must be the smallest of 38/3, 539/39, and 347/29, i.e. 38/3. This produces the solution.

	320	165	190
15	15		
14	4/3		$12\frac{2}{3}$
18		$16\frac{1}{2}$	3/2
LOSS			2/3

The total cost is now 610·5, a reduction of $29{\cdot}5 \times 38/3 = 373\frac{2}{3}$. We compute the new shadow costs from

$$20s_1 + r_1 = 12 \qquad 15s_1 + r_2 = 13 \qquad 14s_3 + r_2 = 11$$
$$10s_2 + r_3 = 15 \qquad 8s_3 + r_3 = 14 \qquad s_3 = 8$$

These equations are solved by

s_1	s_2	s_3	r_1	r_2	r_3
7·6	6·5	8	−140	−101	−50

This produces in the empty cells

$s_i p_{ij} + r_j$

	$i = 1$	2	3
$j = 1$		−23	−44
2		$-16\frac{1}{2}$	
3	56·4		
	7·6	6·5	

Differences

	−35	−60
	−29·5	
38·4		
−7·4	−8·5	

We shall gain by making x_{13} positive. If we make $x_{13} = \theta$, then the total cost will be reduced by $38{\cdot}4\theta$. As a consequence, we have

$$x_{13} = \theta \qquad x_{33} = 3/2 - \theta$$
$$x_{12} = 4/3 - 14\theta/15 \qquad x_{32} = 12\tfrac{2}{3} + 14\theta/15$$
$$x_3 = \tfrac{2}{3} - 76\theta/15$$

To keep the solution basic feasible, we choose θ to be the smallest of 3/2, 5/38 and 10/7, i.e. 5/38. This gives

	320	165	190
15	15		
14	23/19		243/19
18	5/38	$16\frac{1}{2}$	26/19
LOSS			

The total cost is now 605·4, a reduction of 38·4 × 5/38 = 5·1.

To proceed, we compute shadow costs from

$$20s_1 + r_1 = 12 \qquad 10s_2 + r_3 = 15 \qquad 14s_1 + r_3 = 18$$
$$15s_1 + r_2 = 13 \qquad 14s_3 + r_2 = 11 \qquad 8s_3 + r_3 = 14$$

which gives

s_1	s_2	s_3	r_1	r_2	r_3
10/19	83/190	8/19	28/19	97/19	202/19

These shadow costs produce in all empty cells values of $s_i p_{ij} + r_j$ which are smaller than the true costs, so that we have reached the final answer.

The problem of aircraft routing was solved in the way exhibited above in [17]. An equivalent problem arises in connection with machine loading, and was treated in [16], [31], and [3].

5 Critical Path Scheduling

A library has to be transferred from its present position to another one, and on this occasion it has been decided to leave the old shelves behind and to put the books on new ones, in the new location. In order to carry out the whole transfer, the following actions must be taken, and the times these take must be considered:

(*a*)	Order crates and wait for them to be delivered at the present location of the library:	2 days
(*b*)	Order lorries, and wait for them to arrive:	2 days
(*c*)	Choose and buy new shelves; have them delivered at the present location:	2 days
(*d*)	Put books into the crates, and list them:	4 days
(*e*)	Order lorries to take the crates away from the new location, after having been emptied; do not order them before the crates have arrived, because you will then know what lorries you want. They will arrive after:	2 days
(*f*)	Load the crates containing the books on to the lorries:	1 day
(*g*)	Send the book list by post to the new location:	2 days
(*h*)	Put the new shelves on to the lorries:	1 day
(*i*)	Send instructions to the new location about how to put the new shelves up; wait for the shelves to arrive:	1 day
(*j*)	Send the crates and the shelves to the new location, unload:	2 days
(*k*)	Put the shelves up:	2 days
(*l*)	Unpack the books onto the shelves, send the crates away:	6 days

Which is the least time it takes to perform the whole operation?

We start by noticing that certain actions must be finished before certain other actions can start. For instance, the books can not be put into the crates before these have arrived.

To describe the procedure graphically, we draw points 0, 1, 2, . . . for all those times at which some action can start, and draw arrows from points where an action starts to those points where an action can only start when this action has been finished. Thus, for instance, if 3 is the point where action (*g*) can start, and 1 is the point where we can start action (*d*), then we draw an arrow from 1 to 3, because (*g*) can only start when (*d*) has been finished.

We have also to cope with the fact that, for instance, (*g*) can start immediately after (*d*) has been finished, but (*f*) only when both (*b*) and (*d*) have finished, while (*h*) must wait for (*b*) and also for (*c*) to finish. Also, (*i*) has to wait only for (*c*), but not for (*b*) as well. This is the reason for the dotted lines between 2 and 6, 3 and 6, 2 and 5, and between 4 and 5. They represent "activities" whose duration is zero.

In the graph the activities listed above are described by arrows as follows:

(*a*) 0–1	(*b*) 0–2	(*c*) 0–4	(*d*) 1–3
(*e*) 1–10	(*f*) 6–7	(*g*) 3–9	(*h*) 5–7
(*i*) 4–8	(*j*) 7–8	(*k*) 8–9	(*l*) 9–10

We have also indicated the times it takes to perform the various activities which correspond to the several arrows.

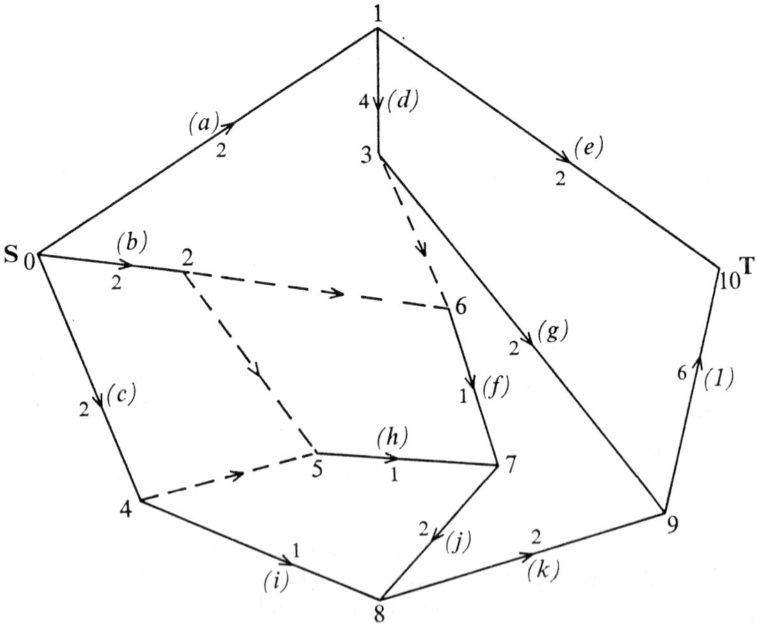

If we look at a particular chain from 0 to 10, say that through the points 1, 3, 9, then we see that the performance of the corresponding activities will take, altogether, $2 + 4 + 2 + 6 = 14$ time units. However, we can not start at point 9 immediately after (g) has been completed because we must wait for the termination of (k) as well, and this can not have happened at time 8. Thus, because we shall have to wait at some points for all the previous activities to have finished, we see that it is the length (in time units) of the longest chain from 0 to 10 which gives the shortest time in which the whole project can be finished.

We are thus led to determining the length of the longest chain between two points, S and T, in a graph of arrows with lengths. We do this by imagining that a traveller walks from 0 to 10 along a succession of links. We describe his path by variables x_{ij} which are equal to 1 if he uses the link from i to j, and equal to zero if he does not.

If the traveller reaches some intermediate point on his path, then he must also leave on some arrow pointing away from that point. Therefore we have for all points except 0 and 10:

$$
\begin{aligned}
&x_{01} - x_{13} - x_{1\,10} = 0 \qquad && x_{02} - x_{25} - x_{26} = 0\\
&x_{13} - x_{36} - x_{39} = 0 && x_{04} - x_{45} - x_{48} = 0\\
&x_{25} + x_{45} - x_{57} = 0 && x_{26} + x_{36} - x_{67} = 0\\
&x_{57} + x_{67} - x_{78} = 0 && x_{48} + x_{78} - x_{89} = 0\\
&x_{39} + x_{89} - x_{9\,10} = 0 &&
\end{aligned}
$$

Moreover, since he must leave 0 along some link, and reach 10 along some link, we have also

$$-x_{01} - x_{02} - x_{04} = -1 \qquad \text{and} \qquad x_{1\,10} + x_{9\,10} = 1$$

The longest chain will be that for which

$$2x_{01} + 2x_{02} + 2x_{04} + 4x_{13} + 2x_{1\,10} + x_{67} + 2x_{39} + x_{57} + x_{48} + 2x_{78} + 2x_{89} + 6x_{9\,10}$$

is maximized.

We have obtained a linear programming problem, with a special structure. To make this structure apparent, we have written the equations in such a way that every variable appears once with a positive coefficient $+1$, and once with a negative coefficient -1. Clearly this is always possible in such a problem.

A problem of this structure can be transformed into one of transportation type. This can be shown by an application of the transhipment device, but in this case it is simplest to introduce it as follows.

For a variable, or a sum of variables with negative sign in one of the first nine equations, introduce the difference between M (a very large constant) and a new variable z_i. If M is large enough, then the z_i will also be non-negative. For instance, for $x_{13} + x_{1\,10}$, which appears in the first equation with negative sign, we write $M - z_1$, and the first equation will then be replaced by the following two:

$$x_{13} + x_{1\,10} + z_1 = M \qquad \text{and} \qquad x_{01} + z_1 = M$$

Treating the other equations in the same fashion, we obtain

$$\begin{aligned}
x_{25} + x_{26} + z_2 &= M & x_{02} + z_2 &= M \\
x_{36} + x_{39} + z_3 &= M & x_{13} + z_3 &= M \\
x_{45} + x_{48} + z_4 &= M & x_{04} + z_4 &= M \\
x_{57} + z_5 &= M & x_{25} + x_{45} + z_5 &= M \\
x_{67} + z_6 &= M & x_{26} + x_{36} + z_6 &= M \\
x_{78} + z_7 &= M & x_{57} + x_{67} + z_7 &= M \\
x_{89} + z_8 &= M & x_{48} + x_{78} + z_8 &= M \\
x_{9\,10} + z_9 &= M & x_{39} + x_{89} + z_9 &= M
\end{aligned}$$

Together with the last two equations, this can now be put into the following table:

	M	M	M	M	M	M	M	M	M	1
M	z_1		(4) x_{13}							(2) $x_{1\,10}$
M		z_2			(0) x_{25}	(0) x_{26}				
M			z_3			(0) x_{36}			(2) x_{39}	
M				z_4	(0) x_{45}			(1) x_{48}		
M					z_5		(1) x_{57}			
M						z_6	(1) x_{67}			
M							z_7	(2) x_{78}		
M								z_8	(2) x_{89}	
M									z_9	(6) $x_{9\,10}$
1	(2) x_{01}	(2) x_{02}		(2) x_{04}						

All columns and rows add up to their marginal values. The latter can be taken to be availabilities and requirements in a transportations scheme, and the costs of this scheme would be

for the x_{ij}	their coefficients in the objective function (given in brackets above their names)
for the z_i	zero
for all cells which are empty in the table	a very large negative (!!) value

These latter cells will remain empty in the final answer, since we want, in this case, to maximize (!!) the total "cost."

The solution can be found by a transportation routine, and the fictitious costs indicated in the table below show that the answer is as follows:

	M	M	M	M	M	M	M	M	M	1	f.c.
M	$M-1$		1								0
M		M									0
M			$M-1$			1					−4
M				M	0						−4
M					M		0				−4
M						$M-1$	1				−4
M							$M-1$	1			−5
M								$M-1$	1		−7
M									$M-1$	1	−9
1	1	0									2
f.c.	0	0	4	4	4	4	5	7	9	15	

A few zeros have been inserted, to indicate a basic solution. All those variables not mentioned in the table are also zero.

The solution means that the longest chain is that from 0 through 1, 3, 6, 7, 8, and 9 to 10, because the variables with value 1 are x_{01}, x_{13}, x_{36}, x_{67}, x_{78}, x_{89}, $x_{9\,10}$. The total length is

$$2 + 4 + 1 + 2 + 2 + 6 = 17$$

This path is called *critical.* Any delay in the activities of its arrows, whether by lengthening the times given, or by not starting at the earliest possible times, would delay the termination beyond time 17.

On those arrows which do not lie on the critical path a delay within limits would not matter. For instance, activity (*b*) (i.e. 02) could start at time 4 rather than time 0, and we could still arrive at point 2, and then at point 6, in time for the activity (*f*), i.e. (67), to start in time on the critical path. Also, we would still arrive in time at point 5, and if we do not further delay the start of (*h*), i.e. (57), we arrive at point 7 without having to delay the start of (*j*), viz. (78).

In order to obtain some information about admissible delays, we formulate the problem of the earliest termination in a new way, and this will not be transformable into a transportation problem.

We call t_i the time at which we can start the activities from point *i*, and make $t_0 = 0$. We have then (meaning by t_{10} the time at which the process could finish):

$$
\begin{array}{lll}
t_1 \geqslant 2 & t_2 \geqslant 2 & t_4 \geqslant 2 \\
t_3 - t_1 \geqslant 4 & t_{10} - t_1 \geqslant 2 & t_7 - t_3 \geqslant 1 \\
t_9 - t_3 \geqslant 2 & t_7 - t_2 \geqslant 1 & t_7 - t_4 \geqslant 1 \\
t_8 - t_4 \geqslant 1 & t_8 - t_7 \geqslant 2 & t_9 - t_8 \geqslant 2 \\
 & & t_{10} - t_9 \geqslant 6
\end{array}
$$

Minimize t_{10}.

It should be noticed how we dealt with the dotted lines on the graph. In particular, $t_7 - t_2 \geqslant 2$ stands for the links 26 and 67, and also for the links 25 and 57. If the lengths of the links 57 and 67 were not the same, then we should have to make the right-hand side of the last inequality equal to the larger of the two.

To solve the linear programming problem, we divide the graph into three portions, by ignoring for the time being the links 36, 45, 78, and 89. This produces three unconnected graphs, with the following nodes:

Graph 1	Graph 2	Graph 3
0, 1, 3, 9, 10	0, 2, 5, 6, 7	0, 4, 8

We tackle three linear programming problems, and combine them afterwards into the whole problem.

The slack variables of an inequality $t_i - t_j \geqslant b_{ij}$ will be denoted by z_{ji}; those of an inequality $t_i \geqslant b_i$ by z_{0i}.

In solving the problems connected with the several graphs, we must remember that it is not our immediate aim to maximize a function, but to find a feasible solution of the constraints. The maximizing or minimizing aspect will be considered when we have re-combined the three cases.

We shall also make use of the observation that, if there are two columns in the tableau such that they do not both have positive entries in the same row, then we can have simultaneously two pivots, one in either column. (See the columns of t_3 and t_{10} in the first tableau below.)

We start with the first graph, concerning t_1, t_3, t_9, and t_{10}.

	t_1	t_3	t_9	t_{10}	
z_{01}	−1				−2
z_{13}	1	−1*			−4
$z_{1\,10}$	1			−1	−2
z_{39}		1	−1		−2
$z_{9\,10}$			1	−1*	−6

	t_1	z_{13}	t_9	$z_{9\,10}$	
z_{01}	−1				−2
t_3	−1	−1			4
$z_{1\,10}$	1		−1	−1	4
z_{39}	1	1	−1*		−6
t_{10}			−1	−1	6

	t_1	z_{13}	z_{39}	$z_{9\,10}$	
z_{01}	−1*				−2
t_3	−1	−1			4
$z_{1\,10}$		−1	−1	−1	10
t_9	−1	−1	−1		6
t_{10}	−1	−1	−1	−1	12

	z_{01}	z_{13}	z_{39}	$z_{9\,10}$	
t_1	−1				2
t_3	−1	−1			6
$z_{1\,10}$		−1	−1	−1	10
t_9	−1	−1	−1		8
t_{10}	−1	−1	−1	−1	14

The second graph concerns t_2, t_5, t_6, and t_7.

	t_2	t_5	t_6	t_7	
z_{02}	−1*				−2
z_{26}	1		−1		0
z_{25}	1	−1			0
z_{57}		1		−1*	−1
z_{67}			1	−1	−1

	z_{02}	t_5	t_6	z_{57}	
t_2	−1				2
z_{26}	1		−1		−2
z_{25}	1	−1*			−2
t_7		−1		−1	1
z_{67}		−1	1	−1	0

	z_{02}	z_{25}	t_6	z_{57}	
t_2	−1				2
z_{26}	1		−1*		−2
t_5	−1	−1			2
t_7	−1	−1		−1	3
z_{67}	−1	−1	1	−1	2

	z_{02}	z_{25}	z_{26}	z_{57}	
t_2	−1				2
t_6	−1		−1		2
t_5	−1	−1			2
t_7	−1	−1		−1	3
z_{67}	0	−1	1	−1	0

Finally, for the third graph we have

	t_4	t_8	
z_{04}	−1*		−2
z_{48}	1	−1	−1

	z_{04}	t_8	
t_4	−1		2
z_{48}	1	−1	−3

	z_{04}	z_{48}	
t_4	−1		2
t_8	−1	−1	3

The three last tableaux of each graph must now be combined. The easiest way of doing this is to introduce the dual problem of the original linear programming formulation, i.e. that referring to the complete graph.

In that formulation, every inequality refers to a link. In the dual, we have corresponding variables, which we call v, and we attach to them subscripts which identify the corresponding link. We have then the following dual problem (introducing the slack variables $s_1, \ldots, s_{10}$):

$$\begin{array}{llll} v_{01} - v_{13} - v_{1\,10} + s_1 = 0 & & v_{26} - v_{67} + v_{36} + s_6 & = 0 \\ v_{02} - v_{25} - v_{26} + s_2 & = 0 & v_{57} + v_{67} - v_{78} + s_7 & = 0 \\ v_{13} - v_{39} - v_{36} + s_3 & = 0 & v_{48} + v_{78} - v_{89} + s_8 & = 0 \\ v_{04} - v_{48} - v_{45} + s_4 & = 0 & v_{39} - v_{9\,10} + v_{89} + s_9 & = 0 \\ v_{25} - v_{57} + v_{45} + s_5 & = 0 & v_{1\,10} + v_{9\,10} + s_{10} & = 0 \end{array}$$

These variables refer to all the original inequalities. When we separated the graph into three parts, then the links 36, 45, 78, and 89 were omitted, and hence when we consider the three subgraphs together, and the dual system of their combination, the variables v_{36}, v_{45}, v_{78} and v_{89} will not, at first, appear. The combination of the duals of the three last tableaux obtained above looks as follows (constructed from the primals, reversing the role of rows and columns, and changing the signs):

			2							1					2	2
	s_1	s_3	$v_{1\,10}$	s_9	s_{10}	s_2	s_6	s_5	s_7	v_{67}	s_4	s_8	v_{36}	v_{45}	v_{78}	v_{89}
2 v_{01}	1	1		1	1								−1			1
4 v_{13}		1	1	1	1								−1			1
2 v_{39}			1	1	1											1
6 $v_{9\,10}$			1		1											
2 v_{02}						1	1	1	1				1	1	−1	
0 v_{25}								1	1	1				1	−1	
0 v_{26}							1			−1			1*			
1 v_{57}									1	1					−1	
2 v_{04}											1	1		−1	1	−1
1 v_{48}												1			1	−1

Ignore, to begin with, the last four columns. Also, we are not interested in the values of the (dual) variables, but rather in the coefficients of the (dual) objective function. These can be computed by using the original coefficients, which we have attached in the table above to the v_{ij} variables in the usual way. We obtain thus the following last row of the tableau:

2 6 10 8 14 | 2 2 2 3 0 | 2 3 || −4 0 −2 3

Now we must see how the last four columns come about. They refer to the four variables which must be introduced to connect again the complete graph. If we look at the constraints of the dual problem, we may say that the introduction of these variables is equivalent to the replacement of

s_3 by $s_3 - v_{36}$ $\quad$ s_4 by $s_4 - v_{45}$
s_5 by $s_5 + v_{45}$ $\quad$ s_6 by $s_6 + v_{36}$
s_7 by $s_7 - v_{78}$ $\quad$ s_8 by $s_8 + v_{78} - v_{89}$

and, finally,

s_9 by $s_9 + v_{89}$

We construct, accordingly, the columns of the new variables $v_{36}, \ldots$ in the following way: v_{36} gets in each row the negative entry of s_3 in the same row, and the positive entry of s_6 in the same row. For instance, in the first row, we write -1 because there is a 1 in the s_3 column. In the v_{26} row, there is a 1 in the s_6 column and hence we write a 1 in the same row, in the v_{36} column. The other new columns are treated in the same way.

The combination of the results for the three subgraphs does not produce the optimum for the complete graph, as we see from the fact that there are now negative values in the bottom row (and we want to maximize in the dual!). We continue by the Simplex Method.

We have to exchange v_{26} for v_{36} (the pivot is indicated above), then v_{57} for v_{67}, and v_{48} for v_{78}. Throughout all these iterations the value of the objective function remains at 14. Finally, we exchange v_{39} for v_{89} and obtain:

	s_1	s_3	$v_{1\,10}$	s_9	s_{10}	s_2	s_6	s_5	s_7	v_{57}	s_4	s_8	v_{26}	v_{45}	v_{48}	v_{89}	
v_{01}	1	1		1	1		1		1	1		1	1		1		1
v_{13}		1	1	1	1		1		1	1		1	1		1		1
v_{89}			1	1	1											1	1
$v_{9\,10}$			1		1												1
v_{02}						1		1		−1			−1	1			0
v_{25}								1		−1				1			0
v_{36}							1		1	1		1	1		1	1	1
v_{67}									1	1		1			1	1	1
v_{04}											1			−1	−1		0
v_{78}												1			1	1	1
	2	6	13	11	17	2	6	2	7	4	2	9	4	0	6	3	17

The intersection of row v_{25} and column v_{45} could be a pivot, but it would not change the values of the variables.

The tableau shows, once more, that t_{10} can not be less than 17. The values of the variables of the primal problem—the one that interests us—are given in the bottom row, so that $t_1 = 2$, $t_3 = 6$, etc., $u_{1\,10} = 13$, and so on. The primal tableau is obtained from the dual tableau above by transposition, and by changing the signs of the entries. We see then that all the t_i have only 0 or -1 in their rows. Hence any increase of the slack variables which have value zero would not decrease any of the t_i: they are the earliest possible ones.

The treatment of critical path problems by methods of linear programming is due to Kelley [27], and was continued in [10]. The dual Simplex method, which we have used above, is due to Lemke [30].

6 Nutrition Problem

The next example illustrates one of the earliest problems in the history of linear programming.

McCance and Widdowson, in *Chemical Composition of Foods*, 2nd ed. (Med. Res. C., 1946), quote the following contents of certain food items:

	Grammes in one ounce of the food item		
	Proteins	Fats	Carbohydrates
1. National wheatmeal bread	2·4	0·3	15·8
2. Cheddar cheese	7·1	9·8	—
3. Roast chicken	8·4	2·1	—
4. Steamed fresh haddock	6·2	0·2	—
5. Dried raw prunes	0·7	—	11·4
6. Walnut (weighed with shells)	2·3	9·4	0·9
7. Gingerbread	1·6	3·7	18·0
8. Margarine	—	28·1	—

The prices of one ounce of these items will be assumed as follows:

Item	1	2	3	4	5	6	7	8
Price in pence	0·5	1·7	3·75	1·0	1·5	1·5	1·75	1·125

It is known that for a grown-up person it is necessary to consume daily at least 75 grammes of proteins, 90 g of fats and 300 g of carbohydrates. If we have to provide these ingredients at the smallest total cost from the food items listed above, what must we buy?

We denote the number of ounces of item i to be bought by x_i, and the number of grammes of ingredient j (= 1 for protein, 2 for fats and 3 for carbohydrates) in item i by a_{ij}. The requirements are then as follows:

$$\begin{aligned}
2{\cdot}4x_1 + 7{\cdot}1x_2 + 8{\cdot}4x_3 + 6{\cdot}2x_4 + 0{\cdot}7x_5 + 2{\cdot}3x_6 + 1{\cdot}6x_7 &\geqslant 75\\
0{\cdot}3x_1 + 9{\cdot}8x_2 + 2{\cdot}1x_3 + 0{\cdot}2x_4 + 9{\cdot}4x_6 + 3{\cdot}7x_7 + 28{\cdot}1x_8 &\geqslant 90\\
15{\cdot}8x_1 + 11{\cdot}4x_5 + 0{\cdot}9x_6 + 18{\cdot}0x_7 &\geqslant 300
\end{aligned}$$

Minimize

$$0{\cdot}5x_1 + 1{\cdot}7x_2 + 3{\cdot}75x_3 + x_4 + 1{\cdot}5x_5 + 1{\cdot}5x_6 + 1{\cdot}75x_7 + 1{\cdot}125x_8$$

We have written the requirements as inequalities, because it might be cheaper to buy an excess of some ingredient, if this is supplied by an item which is cheap and supplies an ample amount of other ingredients as well. The inequalities are transformed into equations by subtracting from the l.h.s. x_9, x_{10} and x_{11} respectively. Inspecting these equations, we find a first feasible solution by the following argument.

We take a large enough supply of prunes (item 5) to supply all carbohydrates, and also some proteins. The remaining proteins can be supplied by roast chicken (item 3), which does not add to the carbohydrates, but also supplies some fats. The remainder of these can be made up by margarine (item 8), which does not add anything to the other two ingredients.

We obtain in this way a feasible solution, but we do not know yet whether it is the cheapest. Therefore we solve for x_3, x_5 and x_8 and apply our tableau technique.

The first tableau shows that the cost of the first solution is 67·770d., i.e. 5s. 8d. As we proceed, we find that this can eventually be reduced to 17·582d., i.e. 1s. 6d. Under the given conditions, and with the available items, this is the best we can do.

In the present case none of the slack variables is contained in a basis. However, if only items 2, 3, and 5 were available, then the constraints would read

$$\begin{aligned}
7{\cdot}1x_2 + 8{\cdot}4x_3 + 0{\cdot}7x_5 - y_1 &= 75\\
9{\cdot}8x_2 + 2{\cdot}1x_3 - y_2 &= 90\\
11{\cdot}4x_5 - y_3 &= 300
\end{aligned}$$

Minimize

$$1{\cdot}7x_2 + 3{\cdot}75x_3 + 1{\cdot}5x_5 = C, \text{ say.}$$

		0·5	1·7	1·0	1·5	1·75			
		x_1	x_2	x_4	x_6	x_7	x_9	x_{10}	x_{11}
3·75 x_3	6·736	0·170	0·845	0·738	0·267	0·059	0·119		0·007
1·5 x_5	26·316	1·386*			0·079	1·579			−0·088
1·125x_8	2·699	−0·002	0·286	−0·048	0·315	0·127	0·009	−0·036	−0·001
	67·770	2·214	1·790	1·714	−0·026	0·983	−0·436	−0·040	−0·107

		1·5	1·7	1·0	1·5	1·75			
		x_5	x_2	x_4	x_6	x_7	x_9	x_{10}	x_{11}
3·75 x_3	3·508	−0·123	0·845	0·738*	0·257	−0·135	−0·119		0·018
0·5 x_1	18·987	0·721			0·057	1·139			−0·063
1·125x_8	2·737	0·001	0·286	−0·048	0·315	0·129	0·009	−0·036	−0·001
	25·728	−1·600	1·790	1·714	−0·153	−1·542	−0·436	−0·040	0·035

		1·5	1·7	3·75	1·5	1·75			
		x_5	x_2	x_3	x_6	x_7	x_9	x_{10}	x_{11}
1·0 x_4	4·753	−0·167	1·145	1·355	0·348	−0·183	−0·161		0·024
0·5 x_1	18·987	0·721			0·057	1·139			−0·063
1·125x_8	2·965	−0·007	0·341	0·065	0·332	0·120	0·001	−0·036	0·000
	17·582	−1·314	−0·171	−2·322	−0·750	−1·228	−0·160	−0·040	−0·007

Note: In the second tableau above, x_4 rather than x_2 was chosen for the new basic variable, because 1·790 is not much larger than 1·714, and a preliminary computation showed that by this choice the reduction would be larger than otherwise.

They can be solved as follows:

$$
\begin{aligned}
x_2 &= 9{\cdot}184 - 0{\cdot}214x_3 + 0{\cdot}102y_2 \\
x_5 &= 26{\cdot}316 + 0{\cdot}088y_3 \\
y_1 &= 8{\cdot}625 + 6{\cdot}878x_3 + 0{\cdot}724y_2 + 0{\cdot}061y_3 \\
C &= 55{\cdot}087 + 3{\cdot}386x_3 + 0{\cdot}173y_2 + 0{\cdot}132y_3
\end{aligned}
$$

The last expression shows that the minimum is 55·087 (of course this is more than before, because the choice is now more restricted). In this case we have to use an oversupply of proteins, as indicated by the positive value of y_1. The constraints without slack variables do not have any feasible solution at all.

However, even when a precise supply of the necessary ingredients, but not more, is feasible it may still be cheaper to buy an oversupply. Such an example is the following, based on a paper by Stigler [34].

He quotes the following necessary daily allowance of nutrients for a moderately active man (weighing 154 pounds):

Calories	3,000
Protein	70 grammes
Calcium . . .	0·8 grammes
Iron	12 milligrammes
Vitamin A . . .	5,000 intern. units
Thiamine (B) . . .	1·8 milligrammes
Riboflavin (B_2 or G) .	2·7 milligrammes
Niacin (nicotinic acid) .	18 milligrammes
Ascorbic acid (C) . .	75 milligrammes

Stigler also gives a list of nutritive values of certain food items per dollar of expenditure, as on 15th August, 1939. From his list we extract the following 10 items:

	*Food items**									
	1	2	3	4	5	6	7	8	9	10
Calories (in 1,000)	44·7	36·0	8·4	15·7	41·7	2·2	2·6	14·3	1·1	26·9
Proteins (g)	1,411	897	422	661	—	333	125	336	106	1,691
Calcium (g)	2·0	1·7	15·1	1·0	—	0·2	4·0	1·8	—	11·4
Iron (mg)	365	99	9	48	—	139	36	118	138	792
Vitamin A (1,000 i.u.)	—	30·9	26·0	—	0·2	169·2	7·2	6·7	918·4	—
Thiamine (mg)	55·4	17·4	3·0	9·6	—	6·4	9·0	29·4	5·7	38·4
Riboflavin (mg)	33·3	7·9	23·5	8·1	0·5	50·8	4·5	7·1	13·8	24·6
Niacin (mg)	441	106	11	471	5	316	26	198	33	217
Ascorbic acid (mg)	—	—	60	—	—	525	5,369	2,522	2,755	—

(Since the units in which the food items are given are those whose price is one dollar, the expression to be minimized is simply the total of all variables.)

* Food items

1 enriched wheat flour
2 corn meal
3 evaporated (canned) milk
4 peanut butter
5 lard
6 beef liver
7 cabbage
8 potatoes
9 spinach
10 dried navy beans

Stigler gave altogether 77 items and proceeded to establish the cheapest supply by trial and error. We quote from his paper (p. 310):

> ". . . . the procedure is experimental because there does not appear to be any direct method of finding the minimum of a linear function subject to linear conditions."

This was, of course, quite true in 1945. Actually, he obtained an answer which, after only one further iteration, leads to the optimum. (I am grateful to Professor G. B. Dantzig for having put at my disposal the results of calculations performed, under his supervision, at the RAND Corporation in Sta Monica, California. The following remarks are based on this source.)

There exists an optimal answer to the problem of supplying the required ingredients precisely, and another answer when over-supply is admitted. The price of the latter is, of course, lower. No food items other than those mentioned above appear in either basic optimal solution, and it follows that when the food items are restricted to these 10, the solution is the same as that of the original nutrition problem. We quote (from Dantzig's results) the following:

Item	*Basic optimal solutions when constraints are*	
	equations	*inequalities*
1	0·0181	0·0295
2	0·0170	
3	0·0452	
4	0·0032	
5	0·0244	
6	0·0153	0·0019
7	0·0092	0·0112
8	0·0051	
9	0·0007	0·0050
10		0·0610
Total cost	0·1382	0·1086

When slack variables are introduced (in the case to which the last column refers), they have the following values in the final solution:

Slack variables for

Protein	Iron	Thiamine	Niacin
77·4135	48·4669	2·3204	9·3160

The cost quoted is that for one day. For a year of 365 days we obtain 50·443 and 39·639 respectively. It is of interest that Stigler's result, arrived at by trial and error, gave a total yearly expenditure of \$39·93, which is very near to the true minimum. He used evaporated milk, but no beef liver, and his solution contained an oversupply (positive slack variables) for the same ingredients as Dantzig's.

7 Blending Aviation Gasolines

An oil company produces the following components:

Alkalytes (A), catalytically cracked gasoline (C), straight run gasoline mixture (S) and Isopentane (I). From these, they obtain the following marketable products by blending—Aviation gasoline of performance number 91 (N), and of performance number 100 (Q), and premium automotive gasoline (Z). The price at which the company sells is 5 (in some units) for one unit of N, 6 for one of Q, and 4 for one of Z.

The following technological conditions apply:

1. The performance number of a mixture of A, C, S and I is obtained by giving weights 107, 93, 87, and 108 to these components, respectively.
2. The vapour pressure of a mixture of A, C, S and I is obtained by giving weights 5, 8, 4 and 20 to these components, respectively. It must not exceed 7 lb/in.^{2}
3. The components A, C, S, and I are produced in 38, 25, 40, and 14 units respectively (1 unit = 100 barrels per calendar day).
4. Z is not subject to any blending conditions, so that all that does not go into either N or Q goes into Z.

With an obvious notation, these conditions can be written as follows:

1. $$107A_N + 93C_N + 87S_N + 108I_N \geqslant 91(A_N + C_N + S_N + I_N)$$
$$107A_Q + 93C_Q + 87S_Q + 108I_Q \geqslant 100(A_Q + C_Q + S_Q + I_Q)$$

2. $$5A_N + 8C_N + 4S_N + 20I_N \leqslant 7(A_N + C_N + S_N + I_N)$$
$$5A_Q + 8C_Q + 4S_Q + 20I_Q \leqslant 7(A_Q + C_Q + S_Q + I_Q)$$

3. $$A_N + A_Q + A_Z = 38$$
$$C_N + C_Q + C_Z = 25$$
$$S_N + S_Q + S_Z = 25$$
$$I_N + I_Q + I_Z = 14$$

4. $$A_Z + C_Z + S_Z + I_Z = 38 + 25 + 40 + 14 - (A_N + C_N + S_N + I_N) - (A_Q + C_Q + S_Q + I_Q)$$

The expression to be maximized is

$$B' = 5(A_N + C_N + S_N + I_N) + 6(A_Q + C_Q + S_Q + I_Q) + 4(A_Z + C_Z + S_Z + I_Z)$$

Substituting (4) into B', replacing the inequalities by equations, and reducing, we obtain

$$\begin{aligned}
-16A_N - 2C_N + 4S_N - 17I_N + X_1 &= 0\\
-7A_Q + 7C_Q + 13S_Q - 8I_Q + X_2 &= 0\\
-2A_N + C_N - 3S_N + 13I_N + X_3 &= 0\\
-2A_Q + C_Q - 3S_Q + 13I_Q + X_4 &= 0
\end{aligned}$$

and the four equations in (3) above.

All variables to be non-negative.

Maximize

$$A_N + C_N + S_N + I_N + 2(A_Q + C_Q + S_Q + I_Q) + 468 = B'$$

These are eight equations for 16 unknowns.

We can ignore the constant 468 in B', provided we add it again to the total price at the end of the calculations. We denote $(B' - 468)$ by B.

It is clear from an inspection of the equations that we can choose as the first basic variables A_Z, C_Z, S_Z, I_Z, and X_1, X_2, X_3, X_4. (The case is degenerate.) This gives the following tableau:

		1	1	1	1	2	2	2	2
		A_N	C_N	S_N	I_N	A_Q	C_Q	S_Q	I_Q
X_1	0	−16	−2	4	−17	0	0	0	0
X_2	0	0	0	0	0	−7	7	13	−8
X_3	0	−2	1	−3	13	0	0	0	0
X_4	0	0	0	0	0	−2	1	−3	13
A_Z	38	1	0	0	0	1	0	0	0
C_Z	25	0	1	0	0	0	1	0	0
S_Z	40	0	0	1	0	0	0	1	0
I_Z	14	0	0	0	1	0	0	0	1
B	0	−1	−1	−1	−1	−2	−2	−2	−2

Proceeding in the usual way, we obtain tableaux with the bases

$$X_1,\ X_2,\ X_3,\ X_4,\ A_Q,\ C_Z,\ S_Z,\ I_Z;$$
$$X_1,\ X_2,\ X_3,\ X_4,\ A_Q,\ C_Q,\ S_Z,\ I_Z;$$
$$X_1,\ S_Q,\ X_3,\ X_4,\ A_Q,\ C_Q,\ S_Z,\ I_Z;$$

and then

TABLEAU A

			1	1	1	1				
			A_N	C_N	S_N	I_N	A_Z	C_Z	X_2	X_4
	X_1	0	−16	−2	4	−17	0	0	0	0
2	S_Q	10·972	0·738	−0·683	0	0	0·738	−0·683	0·090	0·055
	X_3	0	−2	1	−3	13	0	0	0	0
2	I_Q	6·455	0·324	−0·234	0	0	0·324	−0·234	0·021	0·090
2	A_Q	38	1	0	0	0	1	0	0	0
2	C_Q	25	0	1	0	0	0	1	0	0
	S_Z	29·028	−0·738	0·683	1	0	−0·738	0·683	−0·090	−0·055
	I_Z	7·545	−0·324	0·234	0	1	−0·324	0·234	−0·021	−0·090
	B	160·855	3·124	−0·834	−1	−1	4·124	0·166	0·221	0·290
		468								
	B'	628·855								

After this, there are tableaux with the bases

$$X_1,\ S_Q,\ I_N,\ I_Q,\ A_Q,\ C_Q,\ S_Z,\ I_Z;$$
$$S_N,\ S_Q,\ I_N,\ I_Q,\ A_Q,\ C_Q,\ S_Z,\ I_Z;$$

and, finally,

TABLEAU B

			S_Z	C_N	X_1	X_3	A_Z	C_Z	X_2	X_4
1	S_N	29·116	1·003	0·713	−0·040	−0·052	−0·740	0·685	−0·090	−0·055
2	S_Q	10·884	−0·003	−0·713	0·040	0·052	0·740	−0·685	0·090	0·055
1	I_N	6·738	0·232	0·247	−0·018	0·054	−0·171	0·158	−0·021	−0·013
2	I_Q	6·416	−0·001	−0·247	0·018	0·023	0·325	−0·234	0·021	0·090
2	A_Q	37·880	−0·004	−0·040	0·054	0·070	1·003	0·003	0·000	0·000
2	C_Q	25·000	0	1·000	0	0	0	1·000	0	0
1	A_N	0·120	0·004	0·040	−0·054	−0·070	−0·003	0·003	0·000	0·000
	I_Z	0·846	−0·231	0	0	−0·077	−0·154	0·077	0	−0·077
	B	196·333	1·222	0	0·111	0·222	3·222	1·000	0·111	0·222
		468								
	B'	664·333								

This is the final tableau, which implies the following schedule of blending:

	N	Q	Z	Total
	(all entries to be divided by 1,000)			
A	120	37,880	—	38,000
C	—	25,000	—	25,000
S	29,116	10,884	—	40,000
I	6,738	6,416	846	14,000
Total	35,974	80,180	846	117,000
Price per unit	5	6	4	
Income	179,870	481,080	3,384	664,334

This is, of course, only one of the optimal solutions, since there is a zero in the bottom row of the last tableau.

One might be interested to know what would happen if the company applied a more ingenuous scheduling principle, by blending as much as possible into the most highly priced product. We can determine this by satisfying the same constraints as before, but maximizing $A_Q + C_Q + S_Q + I_Q$. The answer is easily found from Tableau A. This is equivalent to taking the price of A_N, C_N, S_N and I_N in the objective function equal to zero. In that case the values in the bottom row of Tableau A are all non-negative and hence $2(A_Q + C_Q + S_Q + I_Q)$ has already been found (the factor 2 is irrelevant). There will then be 29·028 left of S and 7·545 of I, and we must try to use these to the best advantage.

After accounting for Q, there is nothing left of either A or C, so that A_Q, A_N, C_Q, C_N are now zero. We can then show that S_Q, S_N, I_Q, and I_N must also be zero, as follows.

Adding the second and fourth constraints we obtain

$$10S_Q + 5I_Q + X_2 + X_4 = 0$$

Solving the first and third constraint for S_N and I_N we have

$$I_N + 3X_1 + 4X_3 = 0$$
$$S_N + 13X_1 + 17X_3 = 0$$

and so these variables must also all be zero. It follows that the remaining amounts of components S and I must be used up by blending them into Z. This gives the following schedule:

	N	*Q*	*Z*	Total
	(all entries to be divided by 1,000)			
A	—	38,000	—	38,000
C	—	25,000	—	25,000
S	—	10,972	29,028	40,000
I	—	6,455	7,545	14,000
Total	—	80,427	36,573	117,000
Price per unit	5	6	4	
Income	—	482,562	146,292	628,854

It appears that the oil company would be well advised to use L.P. rather than a naïve method of scheduling. The total of 628,854 is, of course, a lower bound of the income that can be obtained. An upper bound is given by assuming that the 36,573, which had to be diverted to Z, might have been diverted to N. Then the income would have been $482{,}562 + 36{,}573 \times 5 = 665{,}427$. This amount is too large, but actually nearer to the true maximum than the lower bound.

The treatment above is due to the authors of [11].

8 Urban Development

Four construction schemes are considered for an urban development plan, and it is estimated that their capital cost and their benefit value (in terms of present values) are as given in the following list:

Scheme	To Construct	Capital cost (in 1000)	Benefit (in 1000)
1	Dual carriageway from A to B	1000	850
2	Link from A to the nearest motorway	3000	2500
3	Shopping area at C	2000	2100
4	Link from C to the nearest motorway	2500	2750

It is also known that if schemes 1 and 2 are both realized, then there would be an extra benefit of 100,000, and if schemes 2 and 4 are both realized, there would be saving in capital cost of 1,000,000.

It would be useless to carry out scheme 3, without having scheme 4 put into operation as well.

The authority responsible for the plan has not more than 5,000,000 at its disposal, and it is required to choose a combination of schemes which gives the highest total benefit, but remaining within the budget restriction.

To formulate this problem mathematically, we introduce the following variables: we let x_i $(i = 1, 2, 3, 4)$ be 1, if scheme i is to be carried out, and otherwise we let it be 0. But in this form this does not deal with the situation when schemes 1 and 2, or schemes 2 and 4 are carried out together. We re-define therefore.

x_1 to be 1 only if scheme 2 is not also implemented.
x_2 to be 1 only if neither scheme 1 nor scheme 4 is to be implemented.
x_4 to be 1 only if scheme 2 is not also implemented.

Furthermore, we introduce x_{12} and x_{24} to be 1 if both schemes to which their subscripts refer are carried out, and to be 0 otherwise.

We have then to maximize the benefit

$$850x_1 + 2500x_2 + 2100x_3 + 2750x_4 + 3450x_{12} + 5250x_{24}$$

subject to the budgetary restriction

$$1000x_1 + 3000x_2 + 2000x_3 + 2500x_4 + 4000x_{12} + 4500x_{24} \leqslant 5000$$

However, this is correct only if we exclude the possibility of both x_1 and x_2 being unity, because in that case we would rather have $x_{12} = 1$, and $x_1 = x_2 = 0$, and similarly for x_2, x_4 and x_{24}.

This can be achieved by the constraints

$$x_1 + x_2 + x_{12} \leqslant 1 \qquad x_2 + x_4 + x_{24} \leqslant 1$$

which ensures that, at most, one variable on the left-hand side of the inequalities can be 1.

It might appear that the case when schemes 1, 2 and 4 are all carried out must also be considered, and an appropriate modification applied, but this case cannot happen, because its cost would exceed the budget available.

The last condition, that scheme 3 should not be carried out without scheme 4, is expressed by

$$x_3 - x_4 - x_{24} \leqslant 0$$

This has the desired effect that if $x_3 = 1$, then either $x_4 = 1$ or $x_{24} = 1$ is necessary. That scheme 4 should be considered without scheme 3 is not excluded, and this is as it should be by the precise wording of the stipulation (though in reality one might doubt that scheme 4 offers the full benefit if there is no shopping area at C).

We have now a linear expression to be minimized, subject to linear conditions. The variables must be non-negative, as in all linear programming problems, and in addition they must be either 0 or 1. We apply now a method particularly suitable for such 0–1 problems.

For this method to be used it is convenient that all the coefficients in the objective function should be negative. To obtain this, we substitute for those variables with positive coefficients (actually all of them in this example) new variables $y_i = 1 - x_i$. These new variables will again have values 0 or 1, but when interpreting them we must remember that $y_i = 0$ means that scheme i is to be carried out, and $y_i = 1$ means the opposite.

After these substitutions we have to maximize

$$16900 - 850y_1 - 2500y_2 - 2100y_3 - 2750y_4 - 3450y_{12} - 5250y_{24}$$

and the constraints are

$$-12000 + 1000y_1 + 3000y_2 + 2000y_3 + 2500y_4 + 4000y_{12} + 4500y_{24} \geqslant 0$$

$$-2 + y_1 + y_2 + y_{12} \geqslant 0 \qquad -2 + y_2 + y_4 + y_{24} \geqslant 0$$

$$1 + y_3 - y_4 - y_{24} \geqslant 0$$

The objective function would be largest if we could make all y_i zero, but this is impossible because of the constraints. The variable which appears in most constraints with positive coefficients is y_2, and we make it equal to 1. We obtain thereby the objective function.

$y_2 = 1$

$$14400 - 850y_1 - 2100y_3 - 2750y_4 - 3450y_{12} - 5250y_{24}$$

subject to

$$-9000 + 1000y_1 + 2000y_3 + 2500y_4 + 4000y_{12} + 4500y_{24} \geqslant 0$$

$$-1 + y_1 + y_{12} \geqslant 0, \; -1 + y_4 + y_{24} \geqslant 0,$$

$$1 + y_3 - y_4 - y_{24} \geqslant 0$$

$y_{24} = 1$

We are still prevented from putting all remaining variables equal to zero, so we make now also $y_{24} = 1$ and obtain the objective function:

$$9150 - 850y_1 - 2100y_3 - 2750y_4 - 3450y_{12}$$

subject to

$$-4500 + 1000y_1 + 2000y_3 + 2500y_4 + 4000y_{12} \geqslant 0$$

$$-1 + y_1 + y_{12} \geqslant 0, \; y_4 \geqslant 0, \; y_3 - y_4 \geqslant 0$$

$y_{12} = 1$

Further relaxation is achieved by $y_{12} = 1$, to give

$$5700 - 850y_1 - 2100y_3 - 2750y_4$$

subject to

$$-500 + 1000y_1 + 2000y_3 + 2500y_4 \geqslant 0$$

$$y_1 \geqslant 0, \; y_4 \geqslant 0, \; y_3 - y_4 \geqslant 0$$

$y_1 = 1$

We make $y_1 = 1$ (thereby reducing the objective function by less than if we made $y_3 = 1$, or $y_4 = 1$) and obtain 4850 as the value of the objective function, while all constraints are now satisfied.

This does not mean, though, that we have solved our problem. We must work our way back to see if other choices would have been better.

The last choice was $y_1 = 1$. With $y_1 = 0$ we could never have reached 4850, whether we used y_3 or y_4 to satisfy the first constraint. But going further back, we could consider $y_{12} = 0$.

$y_2 = 1 \quad y_{12} = 0$

$y_{24} = 1$

We have then to maximize

$$9150 - 850y_1 - 2100y_3 - 2750y_4$$

subject to

$$-4500 + 1000y_1 + 2000y_3 + 2500y_4 \geqslant 0$$
$$-1 + y_1 \geqslant 0, \ y_4 \geqslant 0, \ y_3 - y_4 \geqslant 0$$

$y_1 = 1$

We see from the second constraint that we must have $y_1 = 1$, and the first constraint compels us then to make y_3 as well as y_4 equal to 1, because otherwise its left-hand side would remain negative. But that makes the objective function equal to 3450, less than we have already obtained.

$y_3 = y_4 = 1$

Again going one step further back, to $y_{24} = 0$ (but still $y_2 = 1$), we have to maximize

$$14400 - 850y_1 - 2100y_3 - 2750y_4 - 3450y_{12}$$

subject to

$$-9000 + 1000y_1 + 2000y_3 + 2500y_4 + 4000y_{12} \geqslant 0$$
$$-1 + y_1 + y_{12} \geqslant 0 \qquad -1 + y_4 \geqslant 0 \qquad 1 + y_3 - y_4 \geqslant 0$$

To satisfy the first constraint, all remaining variables in it must be made 1, and thereby the other constraints are also satisfied. This makes the objective function 5250, the best value so far obtained.

However, we must still test what happens if we go one more step back, starting with $y_2 = 0$. The objective function is then

$$16900 - 850y_1 - 2100y_3 - 2750y_4 - 3450y_{12} - 5250y_{24}$$

and the constraints are

$$-12000 + 1000y_1 + 2000y_3 + 2500y_4 + 4000y_{12} + 4500y_{24} \geqslant 0$$
$$-2 + y_1 + y_{12} \geqslant 0 \qquad -2 + y_4 + y_{24} \geqslant 0$$
$$1 + y_3 - y_4 - y_{24} \geqslant 0$$

Because of the second and the third constraint, all variables except y_3 must be made 1, which leaves the objective function to be 4600 $- 2100y_3$, but whatever the value of y_3, this value is less than 5250, which we have reached above.

Thus the problem is solved. The largest benefit we can get is 5,250,000 by making $y_{24} = 0$ and all other $y_1 = 0$. This means that only $x_{24} = 1$, i.e. schemes 2 and 4 are to be carried out, and no other scheme. The cost of this plan is 4,500,000 (well within the available budget).

We have used the method of Balas [2].

9 The Knapsack Problem

A parcel is to be made up of items of three types. An item of type I (or II, or III) weighs 64 lb (or 60 lb, or 35 lb) each. These items are worth, respectively, 6, 5, and 3 monetary units. The parcel must not weigh more than 215 lb and it is desired to make it up so that the total value of all the items contained in it is as large as possible.

We can formulate this problem as one of integer programming. If we denote the number of items i (1, 2, or 3) by x_i, then we want to

$$\begin{aligned} &\text{maximize} && 6x_1 + 5x_2 + 3x_3 \\ &\text{subject to} && 64x_1 + 60x_2 + 35x_3 \leqslant 215 \end{aligned}$$

with x_i integer and non-negative.

However, there are better methods available in this case, which do not take as long as a solution of this integer linear programming problem.

For instance, we may formulate it as a problem of dynamic programming. List the types in descending order of their values. Let a_j be the weight and b_j the value of an item of type j. Denote by $F_k(w)$ the highest total value obtainable by selecting from amongst items of the first k types, when the total weight does not exceed w. Then $F_{k+1}(w)$ is the largest of the values

$$rb_{k+1} + F_k(w - ra_{k+1})$$

where r is one of the numbers 0, 1, 2, . . ., $[w/a_{k+1}]$, where the square brackets indicate the largest integer not exceeding the value they enclose.

To solve our example, we start with $F_1(215) = 18$, obtained by $r_1 = 3$ items of the first type. ($r_1 = 4$ would give a total weight of 256, which would be too large.) We continue by considering

$$F_2(215) = \max_{r_2} [5r_2 + F_1(215 - 60r_2)]$$

so that we must know $F_1(215)$, $F_1(155)$, $F_1(95)$, and $F_1(35)$.

These values are, in the same order, 18, 12, 6 and 0. Of the values

$$0 + 18, \quad 5 + 12, \quad 10 + 6, \quad 15 + 0$$

the first is the largest, and hence $r_2 = 0$; no item of type II will be included at this stage, and we retain the 3 items of type I. We have $F_2(215) = 18$.

Now we consider also the third type, and have

$$F_3(215) = \max_{r_3} [3r_3 + F_2(215 - 35r_3)]$$

so that we must know

$$F_2(215),\quad F_2(180),\quad F_2(145),\quad F_2(110),\quad F_2(75),\quad F_2(40),\quad F_2(5)$$

(i) We already know that $F_2(215) = 18$, obtained by $r_1 = 3$, $r_2 = 0$.

(ii) To find $F_2(180) = \max_{r_2} [5r_2 + F_1(180 - 60r_2)]$

determine first

$$F_1(180) = 12 \quad (\text{with } r_1 = 2)$$
$$F_1(120) = 6 \quad (\text{with } r_1 = 1)$$
$$F_1(60) = F_1(0) = 0 \quad (\text{with } r_1 = 0)$$

and then the largest of the values

$$0 + 12,\quad 5 + 6,\quad 10 + 0,\quad 15 + 0.$$

Of these values the last is the largest, so that

$$F_2(180) = 15 \quad \text{with } r_1 = 0,\quad r_2 = 3$$

(iii) To find $F_2(145) = \max_{r_2} [5r_1 + F_1(145 - 60r_2)]$

determine first

$$F_1(145) = 12 \quad (\text{with } r_1 = 2)$$
$$F_1(85) = 6 \quad (\text{with } r_1 = 1)$$
$$F_1(25) = 0 \quad (\text{with } r_1 = 0)$$

and then the largest of the values

$$0 + 12,\quad 5 + 6,\quad 10 + 0$$

Of these values the first is the largest, so that

$$F_2(145) = 12 \quad \text{with } r_1 = 2,\quad r_2 = 0$$

(iv) To find $F_2(110) = \max_{r_2} [5r_2 + F_1(110 - 60r_2)]$,

determine first

$$F_1(110) = 6 \quad (\text{with } r_1 = 1)$$
$$F_1(50) = 0 \quad (\text{with } r_1 = 0)$$

and then the largest of the values $0 + 6$ and $5 + 0$. The largest is the first, with $r_1 = 1$, $r_2 = 0$, and $F_2(110) = 6$.

Continuing, we have

$$\begin{aligned} F_2(75) &= \max\,[5r_2 + F_1(75 - 60r_2)] \\ &= \max\,[(0 + F_1(75), \quad (5 + F_1(15))] \\ &= \max\,(6, 5), \text{ so that} \end{aligned}$$

(v) $F_2(75) = 6 \quad (r_1 = 1, \quad r_2 = 0)$

(vi) $F_2(40) = 0(r_1 = r_2 = 0)$

(vii) $F_2(5) = 0(r_1 = r_2 = 0)$

We have now to compute

$$F_3(215) = \max$$

$$(0 + 18, \quad 3 + 15, \quad 6 + 12, \quad 9 + 6, \quad 12 + 6, \quad 15 + 0, \quad 18 + 0)$$

The largest value, 18, appears in the first place ($r_1 = 3$, $r_2 = 0$, $r_3 = 0$), in the second ($r_1 = 0$, $r_2 = 3$, $r_3 = 1$), in the third ($r_1 = 2$, $r_2 = 0$, $r_3 = 2$), in the fifth ($r_1 = 1$, $r_2 = 0$, $r_3 = 4$), and in the seventh ($r_1 = 0$, $r_2 = 0$, $r_3 = 6$).

Type I	3	0	2	1	0	
II	0	3	0	0	0	
III	0	1	2	4	6	we call these cases
	a	*b*	*c*	*d*	*e*	← by the letters

Branch and Bound Method

We can also use this example to illustrate the procedure of a "Branch and Bound" Method.

For the time being, ignore the requirement that the numbers of items of the various types must be integers. Then we could, within a total weight of w, accommodate $w/64$ items of type I with a total value of $3w/32$, or $w/60$ items of type II with a total value of $w/12$, or again $w/35$ items of type III with a total value of $3w/35$. Now $3/32 > 3/35 > 1/12$, and we shall consider the types in descending order of these ratios, i.e. in the order I, III, II.

At best, we can, within the restriction to a total weight of not more than 215 lb, obtain a value of $215 \times 3/32 = 20\ 5/32$ (i.e. $20\frac{5}{32}$). With integer values we might not be able to obtain this value, but we shall certainly not obtain any higher one.

At the next step, we investigate the highest value obtainable if 0, 1, 2, or 3 items of type I are packed, assuming unrestricted divisibility of items of the other two types:

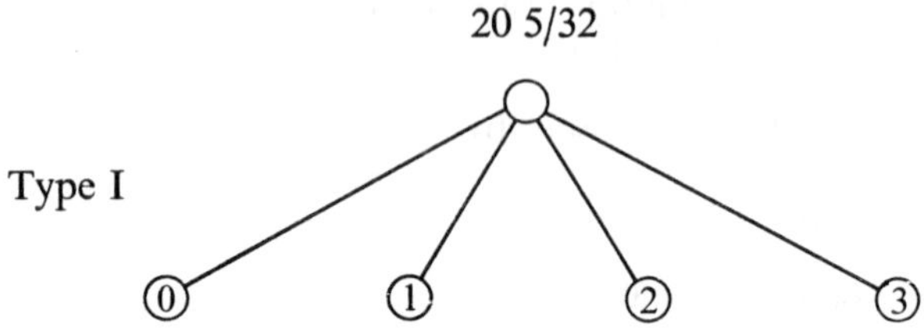

$215 \times 3/35$ $= 18\ 3/7$	$6 + 151 \times 3/35$ $= 18\ 33/35$	$12 + 87 \times 3/35$ $= 19\ 16/35$	$18 + 23 \times 3/35$ $= 19\ 34/35$

The highest of these values is 19 34/35 and, in fact, if 3 items of type I are packed, then no further item of any type can be accommodated. The total value actually obtainable in this manner is 18. This is the case *a* above.

The upper bounds mentioned for the other possibilities are higher than 18 and therefore we must consider these as well. We take the next most promising case, namely that with upper bound 19 16/35, and combine it with various possibilities for items of type III:

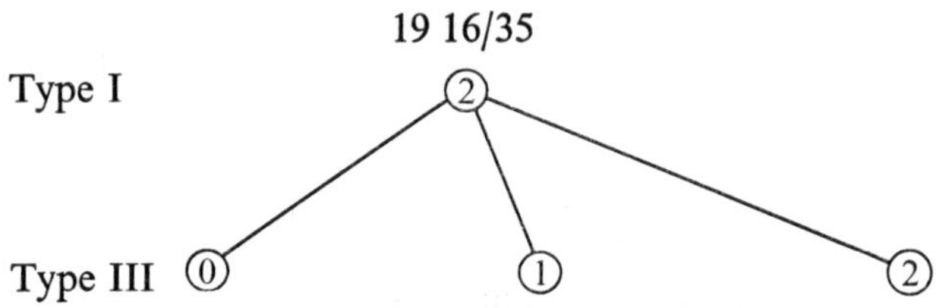

$12 + 87 \times 1/12$ $= 19\ 1/4$	$12 + 3 + 52 \times 1/12$ $= 19\ 1/3$	$12 + 6 + 17 \times 1/12$ $= 19\ 5/12$

The largest values actually obtainable with integers in cases ① and ② for type III are

$$12 + 3 = 15 \qquad \text{and} \qquad 12 + 6 = 18$$

because no complete item of type II could be added. The first of these is less than 18, which we already know is obtainable, and we may ignore this case from now on. The second case gives again 18. It corresponds to case *c* above.

The values where we have left off are

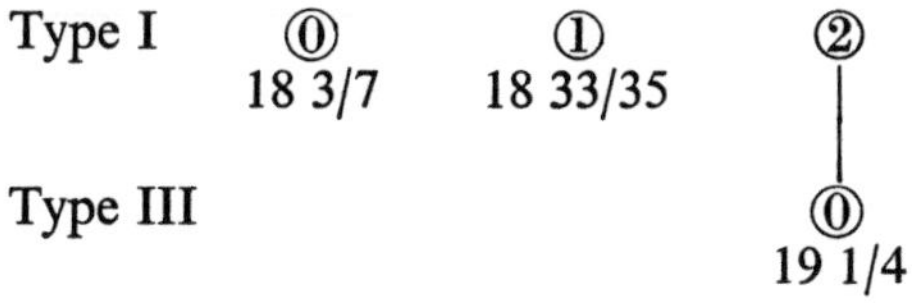

The highest is 19 1/4, and we continue with branches starting from there:

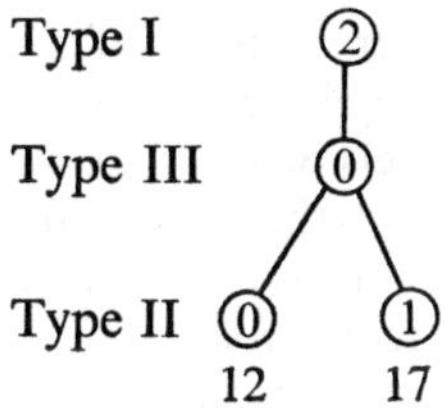

Either value is less than 18 and can therefore be ignored.

We go back to ① of type I and obtain

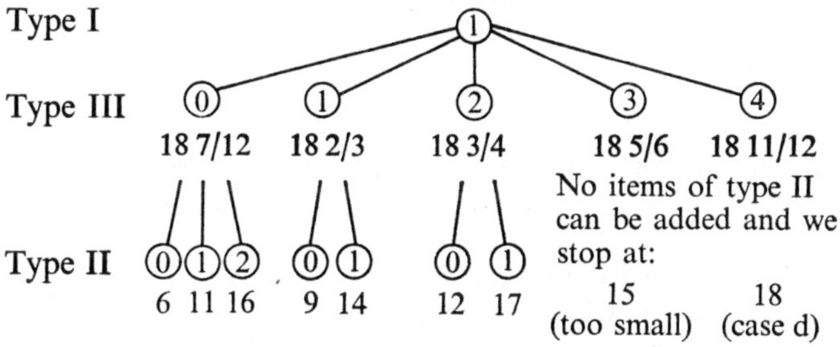

All of these are too small

Finally:

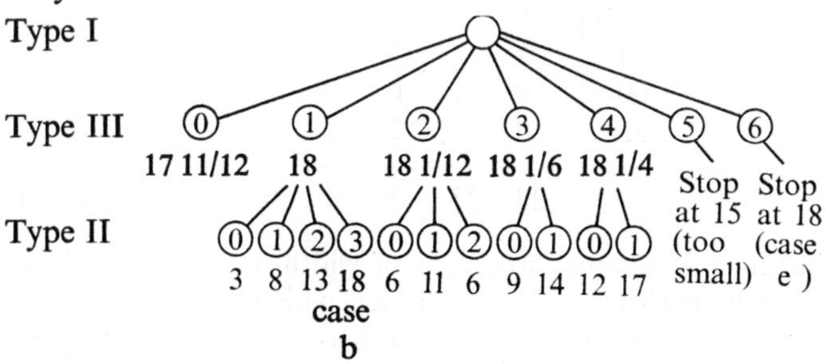

Thus we have again found the five equally good cases *a*, *b*, *c*, *d*, and *e*.

The formulation of the Knapsack problem as one of dynamic programming is mentioned in [12].

10 Trim Loss Reduction

Paper mills produce paper in reels of given width which depends on the manufacturing machine. Customers will require reels of various widths, and the reels of standard width must therefore be cut. As a rule, some wastage at the edge will occur. The manufacturer tries to make the total trim loss as small as possible.

Assume that we have reels of standard width 215 inches, and that the following orders have been received:

Width in inches	*Length ordered*
64	180
60	90
35	90

(As a matter of fact, the orders might be expressed in weights. Conversion into lengths is trivial, if the width is known.)

To begin with, we determine the following possible partitions of 215 inches into the width required:

		Combination $i =$ 1	2	3	4	5	6	7	8	9	10
Widths	64	3	2	2	1	1	1	0	0	0	0
	60	0	1	0	2	1	0	3	2	1	0
	35	0	0	2	0	2	4	1	2	4	6
Trim		23	27	17	31	21	11	0	25	15	5

For instance, if we cut three reels of width 64, no reels of either 60 or 35 can be cut out of the remainder, which is 23.

Denote the length of a standard reel that will be processed according to combination i by x_i. The requirement for a length of 180 and a width of 64 will be satisfied, if

$$3x_1 + 2x_2 + 2x_3 + x_4 + x_5 + x_6 - x_{11} = 180 \qquad \text{(i)}$$

and for the other requirements we need

$$x_2 + 2x_4 + x_5 + 3x_7 + 2x_8 + x_9 - x_{12} = 90 \qquad \text{(ii)}$$

and

$$2x_3 + 2x_5 + 4x_6 + x_7 + 2x_8 + 4x_9 + 6x_{10} - x_{13} = 90 \qquad \text{(iii)}$$

where x_{11}, x_{12}, and x_{13} are slack variables.

Because we require the orders to be met "at least," and not precisely, we do not need to consider combinations other than those quoted. For instance, if a combination with a trim larger than 35 were ever required, then we could just as well say that this is one of the combinations with smaller trim, and one or more of the reels of width 35 included would be in excess of requirements, and hence an addition to the trim loss.

The total trim loss to be minimized is the total length $x_1 + x_2 + \ldots + x_{10}$ times the width 215, less the total area of the paper ordered, i.e. $180 \times 64 + 90 \times 60 + 90 \times 35 = 20{,}070$. Omitting constants, we can therefore say that we want to minimize $x_1 + x_2 + \ldots + x_{10} = C$, say. Then

$$C - x_1 - x_2 - x_3 - x_4 - x_5 - x_6 - x_7 - x_8 - x_9 - x_{10} = 0 \qquad \text{(iv)}$$

We solve this problem by the Inverse Matrix Method.

To find a first feasible solution, we subtract the second equation from the first, and the third equation from twice the first. We also add the first equation to the fourth. We obtain:

$$\begin{array}{l}
3x_1 + 2x_2 + 2x_3 + x_4 + x_5 + x_6 - x_{11} = 180 \\
3x_1 + 2x_2 + 2x_3 - x_4 + x_6 - 3x_7 - 2x_8 - x_9 - x_{11} + x_{12} = 90 \\
6x_1 + 4x_2 + 2x_3 + 2x_4 - 2x_6 - x_7 - 2x_8 - 4x_9 - 6x_{10} - 2x_{11} + x_{13} = 270 \\
C + 2x_1 + x_2 + x_3 - x_7 - x_8 - x_9 - x_{10} - x_{11} = 180
\end{array}$$

The first matrix of the I.M.M. is then

	Constants					x_1	Modified identity matrix			
x_5	1	0	0	0	180	3	1	−1	0	0
x_{12}	0	1	0	0	90	3*	0	1/3	0	0
x_{13}	0	0	1	0	270	6	0	−2	1	0
C	0	0	0	1	180	2	0	−2/3	0	1

We continue by introducing the variable whose coefficient in (4) is largest, viz. x_1, with the pivot in the second row, and hence we pre-multiply the first matrix by the modified identity matrix, given above, to obtain:

x_5	1	-1	0	0	90
x_1	0	1/3	0	0	30
x_{13}	0	-2	1	0	90
C	0	$-2/3$	0	1	120

Using the I.M.M., we would have to find the complete C-row and inspect it to find positive values, if any. However, the number of columns—i.e. the number of possible cutting combinations—might be very large, and we introduce a more convenient method.

We are, in fact, only interested to know whether there is any positive value in the C-row, and this is equivalent to asking whether the largest value there is positive or not.

Denote the coefficients of any variable in the original four equations (i) to (iv) by

a_1
a_2
a_3
$a_4 \ (= -1)$

They must satisfy the condition

$$64a_1 + 60a_2 + 35a_3 \leqslant 215 \qquad (a)$$

The coefficients of the same variable in the equation (1) to (4) are

a_1
$a_1 - a_2$
$2a_1 - a_3$
$a_1 + a_4 \ (= a_1 - 1)$

We are, therefore, since the bottom row of the inverse matrix is now

$$0 - 2/3 \quad 0 \quad 1$$

looking for the maximum of

$$-\tfrac{2}{3}(a_1 - a_2) + a_1 - 1 = \tfrac{1}{3}(a_1 + 2a_2) - 1$$

subject to the inequality (a) above.

The a_i must all be non-negative integers, and we are thus led to solve the integer linear programming problem:

$$\text{maximize } a_1 + 2a_2$$
$$\text{subject to } 64a_1 + 60a_2 + 35a_3 \leqslant 215$$

This simple case can be solved by inspection: $a_1 = 0$, $a_2 = 3$, and $a_3 = 1$ (because we want the slack variable to be smaller than 35).

$\frac{1}{3}(a_1 + 2a_2) - 1$ is then indeed positive, and we introduce the variable corresponding to the combination found. We have

$$\begin{aligned} a_1 &= 0 \\ a_1 - a_2 &= -3 \\ 2a_1 - a_3 &= -1 \\ a_1 - 1 &= -1 \end{aligned}$$

Pre-multiplied by the current inverse matrix, this is

$$\begin{matrix} 3 \\ -1 \\ 5^* \\ 1 \end{matrix} \quad \begin{matrix} \text{and the new modified} \\ \text{identity matrix} \end{matrix} \quad \begin{pmatrix} 1 & 0 & -3/5 & 0 \\ 0 & 1 & 1/5 & 0 \\ 0 & 0 & 1/5 & 0 \\ 0 & 0 & -1/5 & 1 \end{pmatrix}$$

The next inverse matrix is

x_5	1	1/5	−3/5	0	36
x_1	0	−1/15	1/5	0	48
x_7	0	−2/5	1/5	0	18
C	0	−4/15	−1/5	1	102

Searching again for a possible positive value in the C-row, we are looking for the maximum of

$$\begin{aligned} &-4(a_1 - a_2)/15 - (2a_1 - a_3)/12 + a_1 - 1 \\ &= a_1/3 + 4a_2/15 + a_3/5 - 1 \end{aligned}$$

or, equivalently, for the maximum of

$$5a_1 + 4a_2 + 3a_3$$
$$\text{subject to } 64a_1 + 60a_2 + 35a_3 + a_4 = 215$$

(where a_4 is the slack variable). The a_i must be non-negative integers. This is, then, a Knapsack problem and could be solved by the method exhibited in the chapter of that title. We solve it here by an integer linear programming algorithm.

	a_1	a_2	a_3	
a_4	64	60	35*	215
B	−5	−4	−3	0

	a_1	a_2	a_4		
a_3	64/35	12/7	1/35	43/7	introduce a
s_1	−29/35*	−5/7	−1/35	−1/7	further variable
B	17/35	8/7	3/35	129/7	

	s_1	a_2	a_4	
a_3	64/29	4/29	−1/29	169/29
a_1	−35/29	25/29	1/29	5/29
s_2	−23/29*	−25/29	−1/29	−5/29
B	17/29	21/29	2/29	532/29

	s_2	a_2	a_4	
a_3	64/23	−52/23	−3/23	123/23
a_1	−35/23	50/23	2/23	20/23
s_3	−11/23	−4/23	−2/23*	−10/23
B	17/23	2/23	1/23	419/23

	s_2	a_2	s_3	
a_3	7/2	−2	−3/2	6
a_1	−2	2	1	0
a_4	6	2	−23/2	5
B	1/2	0	1/2	18

We could exchange a_1 for a_2, but the answer would remain the same. We find

$$a_1/3 + 4a_2/15 + a_3/5 - 1 = 1/5 > 0$$

and introduce, therefore,

x_{10}							
0	which	18/5*	with the	5/18	0	0	0
0	transforms	−6/5	modified	1/3	1	0	0
6	into	−6/5	identity	1/3	0	1	0
−1		1/5	matrix	−1/18	0	0	1

This leads to

x_{10}	1	1/18	−1/6	0	10
x_1	1/3	0	0	0	60
x_7	1/3	−1/3	0	0	30
C	−1/18	−5/18	−1/6	1	100

Now we have to maximize

$$-a_1/18 - 5(a_1 - a_2)/18 - (2a_1 - a_3)/6 + a_1 - 1$$
$$= a_1/3 + 5a_2/18 + a_3/6 - 1$$

or equivalently, we maximize

$$6a_1 + 5a_2 + 3a_3$$
$$\text{subject to } 64a_1 + 60a_2 + 35a_3 \leqslant 215$$

This is the problem which we have solved in the chapter on the Knapsack problem. There were five solutions, all giving a maximum of 18, and hence a value of 0 to the largest entry in the C-row. We have, therefore, found the minimum of C, which is 100. However, we have more than one single solution.

We quote here the solutions of the Knapsack problem:

a	3	0	2	1	0
a_2	0	3	0	0	0
a_3	0	1	2	4	6

The corresponding variables are here

$x_1 \quad x_7 \quad x_3 \quad x_6 \quad x_{10}$

Now x_1, x_7 and x_{10} are basic, and must therefore by necessity have a 0 in the C-row. But x_{10} could be replaced by x_3 or by x_6. All three are optimal solutions with the same total length and, hence, the same total trim. They are clearly shown in the diagrams.

In this example no slack variables appear in the final solution. The following case shows that it is, in fact, necessary to start off with inequalities, because equations would not have any feasible solution.

Width	Requirements
64	30,000
20	100

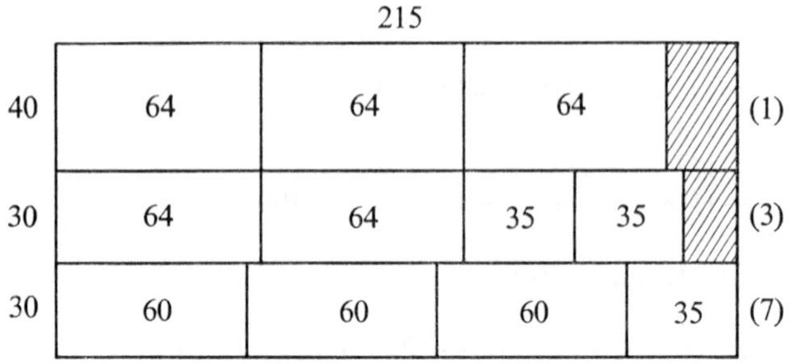

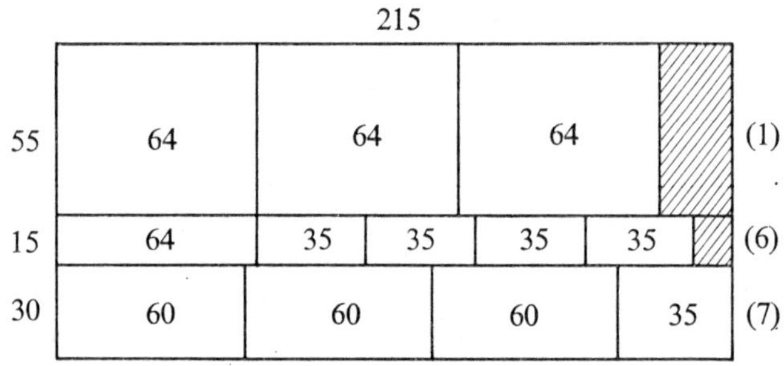

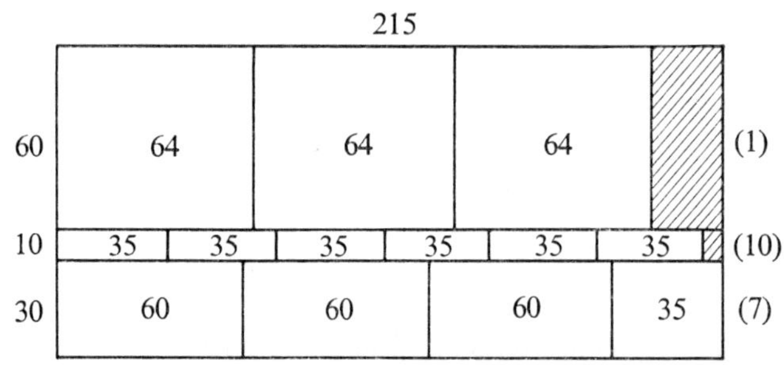

The total length is 100, and the total trim loss is 1430.

Combinations:

	1	2	3	4
64	3	2	1	0
20	1	4	7	10
Trim	3	7	11	15

The equations

$$\begin{aligned} 3x_1 + 2x_2 + x_3 &= 30{,}000 \\ x_1 + 4x_2 + 7x_3 + 10x_4 &= 100 \end{aligned}$$

cannot have feasible solutions, because if we eliminate x_1 we obtain

$$10x_4 + 6\tfrac{2}{3}x_3 + 3\tfrac{1}{3}x_2 = -9{,}900$$

The correct solution to our problem is, of course, $x_1 = 10{,}000$.

We should now like to know how the total trim loss would change if the orders received were different, while the best combination to use remained 1, 7, and 10. We can argue that, since we have already fixed the basic variables to be x_1, x_7 and x_{10}, we can ignore the others and deal with the system:

$$\begin{aligned} 3x_1 \qquad\qquad\qquad &\geqslant 180 \\ 3x_7 \qquad\quad &\geqslant 90 \\ x_7 + 6x_{10} &\geqslant 90 \end{aligned}$$

Minimize $23x_1 + 5x_{10}$.

The dual system is

$$\begin{aligned} 3y_1 \qquad\qquad\qquad &\leqslant 23 \\ 3y_2 + y_3 &\leqslant 0 \\ 6y_3 &\leqslant 5 \end{aligned}$$

Maximize $180y_1 + 90y_2 + 90y_3$.

Because all coefficients in the objective function are positive, it will be maximized if we replace the inequalities by equations, without introducing any slack variables. The solution of the three equations is then

$$y_1 = 23/3, \quad y_2 = -5/18, \quad y_3 = 5/6$$

The trim is 1,430. Hence an addition to the requirements for width 64, 60 or 35 would increase the trim by 23/3, by −5/18, or by 5/6. Because the value of y_2 is negative, it is worth the manufacturer's while to look round for some more orders of width 60. For instance, if we increase the order from 90 to 100, then the solution is $x_1 = 60$, $x_7 = 33\tfrac{1}{3}$, and $x_{10} = 9\tfrac{4}{9}$. The length has been increased to $102\tfrac{7}{9}$, but the trim is now only $60 \times 23 + 9\tfrac{4}{9} \times 5 = 1{,}427\tfrac{2}{9}$, i.e. it has been reduced by $10 \times 5/18$.

Tolerances

Let us now assume that the orders of the customers allow 10% tolerances, i.e. they allow the length ordered to vary within 10% so that the orders are now

Width	Length
64 in.	162–198
60 in.	81–99
35 in.	81–99

In that case a minimization of the trim loss would always lead to the smallest possible length to be delivered. It is more reasonable to demand that the ratio of the wastage to the total length delivered should be minimized.

The wastage due to the trim loss of the various combinations is

$$23x_1 + 27x_2 + 17x_3 + 31x_4 + 21x_5 + 11x_6 \\ + 25x_8 + 15x_9 + 5x_{10}$$

We also have a wastage due to the widths 64, 60, or 35 being cut, in addition to possible deliveries. To deal with these latter wastages, we write the constraints as follows:

$$\begin{aligned} 3x_1 + 2x_2 + 2x_3 + x_4 + x_5 + x_6 - x_{11} + x_{14} &= 198 \\ x_2 + 2x_4 + x_5 + 3x_7 + 2x_8 + x_9 - x_{12} + x_{15} &= 99 \\ 2x_3 + 2x_5 + 4x_6 + x_7 + 2x_8 + 4x_9 + 6x_{10} - x_{13} + x_{16} &= 99 \end{aligned}$$

and add

$$64x_{11} + 60x_{12} + 35x_{13}$$

to the wastage.

Because we must deliver at least the lower limit of the range, we have the further constraints

$$\begin{aligned} -x_{11} + x_{14} + x_{17} &= 36 \\ -x_{12} + x_{15} + x_{18} &= 18 \\ -x_{13} + x_{16} + x_{19} &= 18 \end{aligned}$$

where x_{17}, x_{18}, and x_{19} are, of course, slack variables.

This formulation is valid because, on minimizing the wastage ratio, x_{11}, x_{12}, and x_{13} will only be positive if x_{14}, x_{15}, and x_{16}, respectively, are zero, so that the former variables mean the real wastage.

We want, then, to minimize

$$\frac{23x_1 + 27x_2 + \ldots + 15x_9 + 5x_{10} + 64x_{11} + 60x_{12} + 35x_{13}}{x_1 + x_2 + x_3 + x_4 + x_5 + x_6 + x_7 + x_8 + x_9 + x_{10}}$$

where all variables must be non-negative, and subject to the six constraints above.

The minimization of a ratio of linear functions, subject to linear constraints (so-called fractional, or hyperbolic, linear programming), can be carried out by a number of algorithms, which have been proposed in the literature. We choose here the procedure of Charnes and Cooper [9].

We note that the denominator of the objective function is, in this example, always positive in the feasible region, defined by the constraints. We multiply numerator and denominator by

$$t = (x_1 + x_2 + x_3 + x_4 + x_5 + x_6 + x_7 + x_8 + x_9 + x_{10})^{-1}$$

and introduce $y_i = tx_i$ ($i = 1, \ldots, 19$). Our problem is then to minimize

$$\begin{aligned}&23y_1 + 27y_2 + 17y_3 + 31y_4 + 21y_5 + 11y_6 + 25y_8 + 15y_9\\ &+ 5y_{10} + 64y_{11} + 60y_{12} + 35y_{13}\\ &+ M(x_{101} + x_{102} + x_{103} + x_{104})\end{aligned}$$

subject to $y_i, t \geqslant 0$, and

$$\begin{aligned}&x_{101} + y_1 + y_2 + y_3 + y_4 + y_5 + y_6 + y_7 + y_8 + y_9 + y_{10}\\ &\quad = 1\\ &x_{102} + 3y_1 + 2y_2 + 2y_3 + y_4 + y_5 + y_6 - y_{11} + y_{14} - 198t\\ &\quad = 0\\ &x_{103} + y_2 + 2y_4 + y_5 + 3y_7 + 2y_8 + y_9 - y_{12} + y_{15} - 99t\\ &\quad = 0\\ &x_{104} + 2y_3 + 2y_5 + 4y_6 + y_7 + 2y_8 + 4y_9 + 6y_{10} - y_{13} + y_{16}\\ &\quad - 99t = 0\\ &-y_{11} + y_{14} + y_{17} - 36t = 0\\ &-y_{12} + y_{15} + y_{18} - 18t = 0\\ &-y_{13} + y_{16} + y_{19} - 18t = 0\end{aligned}$$

where the x_{101}, x_{102}, x_{103}, and x_{104} are artificial variables, to start off with.

We use the Inverse Matrix Method and have, to begin with,

x_{101}	1	0	0	0	0	0	0	0	1
x_{102}	0	1	0	0	0	0	0	0	0
x_{103}	0	0	1	0	0	0	0	0	0
x_{104}	0	0	0	1	0	0	0	0	0
y_{17}	0	0	0	0	1	0	0	0	0
y_{18}	0	0	0	0	0	1	0	0	0
y_{19}	0	0	0	0	0	0	1	0	0
C	0	0	0	0	0	0	0	1	M

We try now to introduce those variables, which we have found to be basic in the final solution of the previous problem, without tolerances. We start with

$$\begin{array}{l} y_1 \\ 1 \\ 3^* \\ 0 \\ 0 \\ 0 \\ 0 \\ 0 \\ 4M - 23 \end{array}$$

which leads to

									Constants	New variable y_7 Originally	New variable y_7 Transformed
x_{101}	1	$-1/3$	0	0	0	0	0	0	1	1	1
y_1	0	$1/3$	0	0	0	0	0	0	0	0	0
x_{103}	0	0	1	0	0	0	0	0	0	3	3*
x_{104}	0	0	0	1	0	0	0	0	0	1	1
y_{17}	0	0	0	0	1	0	0	0	0	0	0
y_{18}	0	0	0	0	0	1	0	0	0	0	0
y_{19}	0	0	0	0	0	0	1	0	0	0	0
C	0	$\frac{-4M+23}{3}$	0	0	0	0	0	1	M	$5M$	$5M$

(The last two columns are identical, because of the 0 in the second row)

									Constants	New variable y_{10} Originally, transformed
x_{101}	1	$-1/3$	$-1/3$	0	0	0	0	0	1	1
y_1	0	$1/3$	0	0	0	0	0	0	0	0
y_7	0	0	$1/3$	0	0	0	0	0	0	0
x_{104}	0	0	$-1/3$	1	0	0	0	0	0	6*
y_{17}	0	0	0	0	1	0	0	0	0	0
y_{18}	0	0	0	0	0	1	0	0	0	0
y_{19}	0	0	0	0	0	0	1	0	0	0
C	0	$\frac{-4M+23}{3}$	$\frac{-5M}{3}$	0	0	0	0	1	M	$7M - 5$

										New variable t Originally	New variable t Transformed
x_{101}	1	−1/3	−5/18	−1/6	0	0	0	0	1	0	110*
y_1	0	1/3	0	0	0	0	0	0	0	−198	−66
y_7	0	0	1/3	0	0	0	0	0	0	−99	−33
y_{10}	0	0	−1/18	1/6	0	0	0	0	0	−99	−11
y_{17}	0	0	0	0	1	0	0	0	0	−36	−36
y_{18}	0	0	0	0	0	1	0	0	0	−18	−18
y_{19}	0	0	0	0	0	0	1	0	0	−18	−18
C	0	$\frac{-4M+23}{3}$	$\frac{-23M-5}{18}$	$\frac{-7M+5}{6}$	0	0	0	1	M	$-396M$	$110M-1573$

Since we know that t must be positive, i.e. basic in the final answer, we introduce it now to get rid of the only artificial variable still in the basis.

									Constants
t	1/110	−1/330	−1/396	−1/660	0	0	0	0	1/110
y_1	3/5	2/15	−1/6	−1/10	0	0	0	0	3/5
y_7	3/10	−1/10	1/4	−1/20	0	0	0	0	3/10
y_{10}	1/10	−1/30	−1/12	3/20	0	0	0	0	1/10
y_{17}	36/110	−12/110	−1/11	−6/110	1	0	0	0	36/110
y_{18}	18/110	−6/110	−5/110	−3/110	0	1	0	0	18/110
y_{19}	18/110	−6/110	−5/110	−3/110	0	0	1	0	18/110
C	$143/10 - M$	$29/10 - M$	$-51/12 - M$	$-31/20 - M$	0	0	0	1	143/10

We investigate now the variables y_{14}, y_{15}, and y_{16} which produce in the C-row, 29/10, −51/12, and −31/20. Because the first is positive, we consider y_{14} in greater detail.

The column of y_{14} is

Originally	Transformed to the present stage
0	−1/330
1	2/15
0	−1/10
0	−1/30
1	98/110*
0	−6/110
0	−6/110
M	29/10

A comparison with the constants in the last tableau shows that the pivot is in the fifth row, so that the next tableau is

t	1/98	−1/294	−5/1764	−1/588	1/294	0	0	0	1/98
y_1	27/49	22/147	−15/98	−9/98	−22/147	0	0	0	27/49
y_7	33/98	−11/98	47/196	−11/196	11/98	0	0	0	33/98
y_{10}	11/98	−11/294	−17/196	29/196	11/294	0	0	0	11/98
y_{14}	18/49	−6/49	−5/49	−3/49	55/49	0	0	0	18/49
y_{18}	9/49	−3/49	−5/98	−3/98	3/49	1	0	0	9/49
y_{19}	9/49	−3/49	−5/98	−3/98	3/49	0	1	0	9/49
C	1297/98	319/98	−775/196	−269/196	−319/98	0	0	1	1297/98

If we now compute the values which the remaining variables have in the C-row—by multiplying the C-row of the last tableau into the original columns of those variables—then we find that the entries in the C-row are negative for y_2, y_4, y_5, y_8, and y_9, so that these variables need not be considered, while the entries for y_3 and y_6 are zero. These variables will therefore appear in alternative optima, and the basis of the last tableau is also optimal. We have thus altogether three optimal solutions, which we quote here:

$t = 1/98$	$t = 1/98$	$t = 1/98$
$y_1 = 27/49$	$y_1 = 32/98$	$y_1 = 97/196$
$y_7 = 33/98$	$y_7 = 33/98$	$y_7 = 33/98$
$y_{10} = 11/98$	$y_3 = 33/98$	$y_6 = 33/196$
$y_{14} = 18/49$	$y_{14} = 18/49$	$y_{14} = 18/49$
$y_{18} = 9/49$	$y_{18} = 9/49$	$y_{18} = 9/49$
$y_{19} = 9/49$	$y_{19} = 9/49$	$y_{19} = 9/49$

$C = 1297/98$ (in all cases)

Dividing the y's by t, we return to the x's. These are the resulting three optimal answers:

$x_1 = 54$	$x_7 = 33$	$x_{10} = 11$
$x_1 = 32$	$x_7 = 33$	$x_3 = 33$
$x_1 = 48{\cdot}5$	$x_7 = 33$	$x_6 = 16{\cdot}5$

Total length cut = 98

It is interesting to compare this result with the earlier one, without tolerances. In that case the total wastage was 1,430, and the length cut was 100. That ratio is larger than the present one, with a wastage of 1,297, and a total length of 98.

The trim problem has been discussed in [15] and in [8], and together with customer tolerances in [13]. The integer programming algorithm used above is taken from [22].

11 Production Scheduling

A manufacturer knows that he must supply the following volumes of his product during the next four months:

i	1st	2nd	3rd	4th	month
r_i	5	6	8	6	unit volumes

The production of one unit costs, during the i-th month,

c_i	1	4	2	4	monetary units in regular time
c_i'	2	6	4	6	monetary units in overtime

There is a ceiling on production in that not more than $B_i = 9$ (for all i) can be produced in regular time, and not more than $B_i' = 3$ in overtime (again for all i).

Storage costs $d = 1$ per unit per month. How much should the manufacturer produce in each month?

Because of the changing cost, and because of the capacity restriction, it might be useful to produce in advance of the month when the requirement arises, having regard to the storage cost. We assume, for simplicity, that an item produced during month i and required during month j incurs a storage cost for $j - i$ months.

We denote the volume of production during the i-th month by x_i (in regular time) and by x_i' (in overtime). Then we have the following constraints:

$$x_i \leqslant B_i \qquad \text{and} \qquad x_i' \leqslant B_i' \qquad (i = 1, 2, 3, 4)$$

$$\sum_{i=1}^{t} (x_i + x_i') \geqslant \sum_{i=1}^{t} r_i \qquad (t = 1, 2, 3, 4)$$

The total cost of production and storage, which we want to minimize, is

$$\sum_{i=1}^{4} (c_i x_i + c_i' x_i') + \sum_{i=1}^{4} s_t$$

where we have written s_t for $\sum_{i=1}^{t} r_i - \sum_{i=1}^{t} (x_i + x_i')$.

We introduce slack variables $B_i - x_i = u_i$, and $B_i' - x_i' = u_i'$ and can then transform our problem into one of transportation type, by the following argument.*

A production of x_i, or x_i', contributes towards the requirements of the i-th and of any following month. The s_t are made up of those items that have been produced, but are not required up to (and including) the t-th month. Hence, if x_{ij} and x_{ij}' are the volumes produced, in regular and overtime respectively, during the i-th month for the requirement of the j-th month, then we have

$$\sum_j x_{ij} = x_i \qquad \text{and} \qquad \sum_j x_{ij}' = x_i'$$

If we remember also that the full capacity might not be necessary to cover the requirements, then we obtain the following table:

	5	6	8	6	23
9	x_{11}	x_{12}	x_{13}	x_{14}	u_1
3	x_{11}'	x_{12}'	x_{13}'	x_{14}'	u_1'
9		x_{22}	x_{23}	x_{24}	u_2
3		x_{22}'	x_{23}'	x_{24}'	u_2'
9			x_{33}	x_{34}	u_3
3			x_{33}'	x_{34}'	u_3'
9				x_{44}	u_4
3				x_{44}'	u_4'

23 is the unused capacity, and the entries must add up to the marginal row and column totals.

Of course, we can never produce for the requirements of earlier months; hence the left-hand lower part of the table must remain empty. We cope with this by introducing here a very high cost, M say, while otherwise the costs of the x_{ij} and x_{ij}' are made up of production costs and storage costs. The costs of the u_i and u_i' are zero.

Cost table:

1	2	3	4	0
2	3	4	5	0
M	4	5	6	0
M	6	7	8	0
M	M	2	3	0
M	M	4	5	0
M	M	M	4	0
M	M	M	6	0

* cf. [6].

The following is easily seen to be the cheapest scheme:

	5	6	8	6	23	fict. c.
9	5	4				0
3		2			1	1
9					9	1
3					3	1
9			8	1		0
3					3	1
9				5	4	1
3					3	1
fict. c.	1	2	2	3	−1	

We should produce

in the first month	9 in regular and 2 in overtime
in the second month	nothing at all
in the third month	9 in regular time only, and
in the fourth month	5 in regular time.

We are never told to produce in overtime before the capacity of regular time production is fully utilized. This is reasonable, and will always be so when overtime is more expensive than regular time.

On the other hand, we might think it undesirable that the production should change in such an erratic way from month to month, as it does here: 11, 0, 9, 5 in the successive months. Therefore we consider now the introduction of a further condition, viz. that production must not decrease from one month to the next, i.e. $x_i \geqslant x_{i-1}$ $(i = 2, 3, 4)$. Also, we assume now that an increase of one unit from one month to the next costs 2, so that we add

$$\sum_{i=2}^{4} 2(x_i - x_{i-1})$$

to the objective function.

We shall now omit the consideration of overtime, since it does not add anything significant to the method of solution.

We introduce $y_i = x_i - x_{i-1}$ $(i = 2, 3, 4)$, a non-negative variable. If x_1 is non-negative, as it must be, then so will x_2, x_3 and

x_4, so that we can replace them in terms of x_1 and the y_1 without losing any constraint. We have then

$$
\begin{aligned}
x_1 \qquad\qquad\qquad\qquad &- s_1 = 5 \\
2x_1 + y_2 \qquad\qquad\quad &- s_2 = 11 \\
3x_1 + 2y_2 + y_3 \qquad &- s_3 = 19 \\
4x_1 + 3y_2 + 2y_3 + y_4 &- s_4 = 25 \\
x_1 + u_1 \qquad\qquad\qquad &= 9 \\
x_1 + y_2 + u_2 \qquad\qquad &= 9 \\
x_1 + y_2 + y_3 + u_3 \qquad &= 9 \\
x_1 + y_2 + y_3 + y_4 + u_4 &= 9
\end{aligned}
$$

and we minimize

$$
\begin{aligned}
&x_1 + 4(x_1 + y_2) + 2(x_1 + y_2 + y_3) + 4(x_1 + y_2 + y_3 + y_4) \\
&+ s_1 + s_2 + s_3 + s_4 + 2(y_2 + y_3 + y_4) \\
&= 11x_1 + 12y_2 + 8y_3 + 6y_4 + s_1 + s_2 + s_3 + s_4
\end{aligned}
$$

This calls for an application of the Dual Simplex Method. We start with

		x_1	y_2	y_3	y_4	
1	s_1	−1				−5
1	s_2	−2	−1			−11
1	s_3	−3*	−2	−1		−19
1	s_4	−4	−3	−2	−1	−25
	u_1	1				9
	u_2	1	1			9
	u_3	1	1	1		9
	u_4	1	1	1	1	9
		−21	−18	−11	−7	−60

and proceed to

	s_3	y_2	y_3	y_4	
s_1	−1/3	2/3	1/3		4/3
s_2	−2/3	1/3	2/3		5/3
x_1	−1/3	2/3	1/3		19/3
s_4	−4/3	−1/3	−2/3	−1	1/3
u_1	1/3	−2/3	−1/3		8/3
u_2	1/3	1/3	−1/3		8/3
u_3	1/3	1/3	2/3		8/3
u_4	1/3	1/3	2/3	1	8/3
	−7	−4	−4	−7	73

We produce 19/3 in every month (no increase ever, because the y_i are all zero). The cost account looks as follows:

	Produce	Cost	Supply	Store	Cost
1st month	19/3	19/3	5	4/3	4/3
2nd month	19/3	76/3	6	5/3	5/3
3rd month	19/3	38/3	8	0	0
4th month	19/3	76/3	6	1/3	1/3
		209/3			10/3
Total cost		219/3 = 73			

If we have to take into account that the storage space is limited, so that not more than 1·5 can be stored per month, then we introduce a new variable $s_i' = 1{\cdot}5 - s_i$, which is also restricted to lie between 0 and 1·5. The only s_i which exceeds 1·5 is now s_2. Its row is

s_2 $-2/3$ $1/3$ $2/3$ 0 $5/3$

which is equivalent to

s_2' $2/3$ $-1/3$ $-2/3^*$ 0 $-1/6$

Continuing, again by the Dual Simplex Method, we have $-2/3$ as the pivot, and the next tableau:

	s_3	y_2	s_2'	y_4	
s_1		1/2	1/2		5/4
y_3	−1	1/2	−3/2		1/4
x_1		1/2	1/2		25/4
s_4	−2		−1	−1	1/2
u_1		−1/2	−1/2		11/4
u_2		1/2	−1/2		11/4
u_3	1		1		5/2
u_4	1		1	1	5/2
	−11	−2	−6	−7	74

This means

Month	Produce	Cost	Supply	Store	Cost
1	25/4	25/4	5	5/4	5/4
2	25/4	25	6	3/2	3/2
3	13/2	$13 + \frac{1}{2}$	8	0	0
4	13/2	26	6	1/2	1/2
		$70\frac{3}{4}$			$3\frac{1}{4}$
Total cost		74			

In the last two cases we notice, perhaps with some surprise, that s_4 is positive. This is the result of the smoothening out of the production. The cost of this storage is included in the total cost, though in reality it would, of course, be avoided.

We see also, from the shadow cost of s_2', that it would pay to buy additional storage space up to a cost of 6 per unit in the second month, but not beyond 5/3, the value of s_2. An increase from 3/2 to 5/3 would cost $6(5/3 - 3/2) = 1$, and this is precisely the difference between the costs, 74 and 73, in the last two tableaux.

12 Use of Raw Material

A cloth manufacturer produces four types of cloth, in various combinations of colour. One unit length of the cloth uses amounts of wool of these colours as given in the following table:

Colour of wool	Cloth of type A	B	C	D	Amount of wool available (in some appropriate unit)
Green	1	2	1	1	10
red	2	1	2	1	6
blue	3	1	0	0	10
yellow	1	4	0	0	18
brown	0	0	1	3	8
purple	0	0	3	3	12

From one unit of cloth, he obtains the following profit (in some monetary unit):

3 5 4 1

How should he use the available material to his best advantage?

The system to be solved is as follows:

$$
\begin{aligned}
&\text{maximize } 3x_1 + 5x_2 + 4x_3 + x_4 = x_0, \text{ say} \\
&\text{subject to } x_i \geqslant 0 \\
&x_1 + 2x_2 + x_3 + x_4 \leqslant 10 \\
&2x_1 + x_2 + 2x_3 + x_4 \leqslant 6 \\
&3x_1 + x_2 \leqslant 10 \\
&x_1 + 4x_2 \leqslant 18 \\
&x_3 + 3x_4 \leqslant 8 \\
&3x_3 + 3x_4 \leqslant 12
\end{aligned}
$$

Its structure suggests that we should use the Decomposition Algorithm.*

* See [12] and literature mentioned there.

We draw here diagrams of the feasible regions when the constraints are

the third and fourth inequalities (BLOCK I) and
the fifth and sixth inequalities (BLOCK II).

The numbers of the vertices are inserted in small circles.

We introduce slack variables x_5 and x_6 in the first two (the "linking") equations and note that the origins are vertices in both diagrams. Therefore we can start with the following matrix:

x_5	1					10
x_6		1				6
λ_0			1			1
μ_0				1		1
x_0					1	0

To deal with block I, we have to minimize $-3x_1 - 5x_2$, subject to the constraints of that block, which gives vertex ②, (2, 4).

At this vertex we have $x_1 + 2x_2 = 10$ and $2x_1 + x_2 = 8$, so that the column of λ_2 is

$$\begin{array}{r} 10 \\ 8^* \\ 1 \\ 0 \\ -26 \end{array}$$

Of the ratios 10/10, 6/8, 1/1, 1/0 the second is smallest, and thus the pivot is in the second row, indicated by an asterisk.

It follows that our next matrix is

x_5	1	$-5/4$	0	0	0	5/2
λ_2	0	1/8	0	0	0	3/4
λ_0	0	$-1/8$	1	0	0	1/4
μ_0	0	0	0	1	0	1
x_0	0	13/4	0	0	1	39/2

To carry on, we have to minimize

$$13(2x_1 + x_2)/4 - (3x_1 + 5x_2) = 7x_1/2 - 7x_2/4$$

and this leads to vertex ③, with the column of λ_3 which was originally

$$\begin{array}{r} 9 \\ 9/2 \\ 1 \\ 0 \\ -22\tfrac{1}{2} \end{array}$$

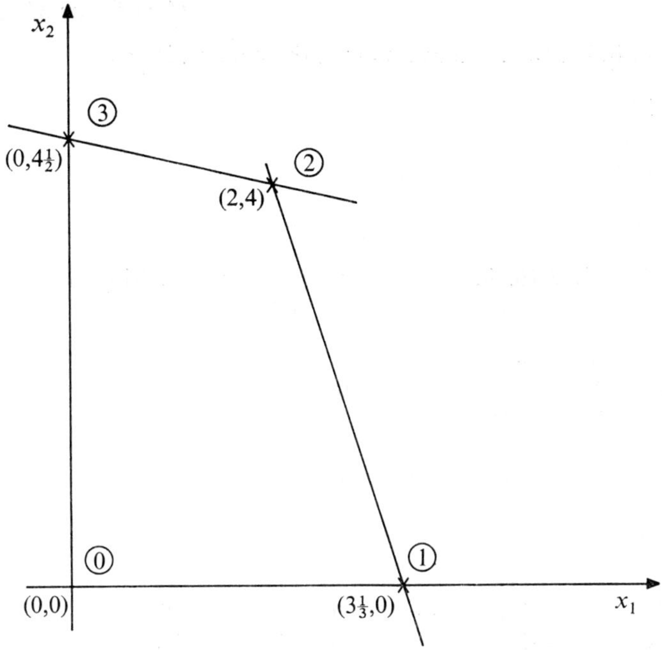

BLOCK I

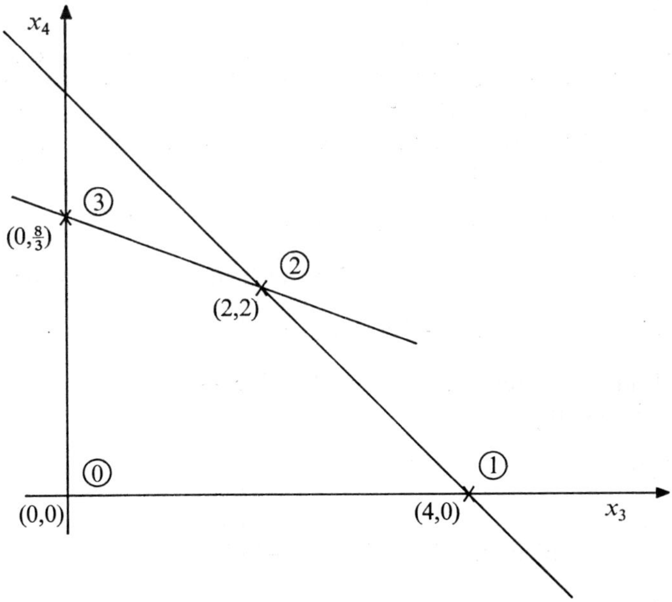

BLOCK II

Pre-multiplied by the above matrix, the column is now

$$\begin{matrix} 27/8 \\ 9/16 \\ 7/16^* \\ 0 \\ -63/8 \end{matrix}$$

The pivot is 7/16, in the third row, and pre-multiplying the last inverse matrix by the modified unit matrix:

$$\begin{matrix} 1 & 0 & -54/7 & 0 & 0 \\ 0 & 1 & -9/7 & 0 & 0 \\ 0 & 0 & 16/7 & 0 & 0 \\ 0 & 0 & 0 & 1 & 0 \\ 0 & 0 & 18 & 0 & 1 \end{matrix}$$

we obtain

$$\begin{array}{cccccc|c} x_5 & 1 & -2/7 & -54/7 & 0 & 0 & 4/7 \\ \lambda_2 & 0 & 2/7 & -9/7 & 0 & 0 & 3/7 \\ \lambda_3 & 0 & -2/7 & 16/7 & 0 & 0 & 4/7 \\ \mu_0 & 0 & 0 & 0 & 1 & 0 & 1 \\ x_0 & 0 & 1 & 18 & 0 & 1 & 24 \end{array}$$

The last column, which gives the present values of the variables, was obtained by pre-multiplying the previous last column by the latest modified unit matrix. As a check, we pre-multiply the original column of the constants by the inverse matrix just obtained.

Continuing, we minimize

$$2x_1 + x_2 + 18 - 3x_1 - 5x_2 = -x_1 - 4x_2 + 18$$

for block I, and obtain 0. Therefore we turn now to block II:

$$\text{minimize } 2x_3 + x_4 - 4x_3 - x_4 = -2x_3$$

This leads to vertex ①, (4, 0), with value −8. We introduce the original column:

$$\begin{matrix} \mu_1 \\ 4 \\ 8 \\ 0 \\ 1 \\ -16 \end{matrix} \qquad \text{which transforms into} \qquad \begin{matrix} 12/7 \\ 16/7^* \\ -16/7 \\ 1 \\ -8 \end{matrix}$$

with the pivot in the second row. Hence we pre-multiply by

1	−3/4	0	0	0
0	7/16	0	0	0
0	1	1	0	0
0	−7/16	0	1	0
0	7/2	0	0	1

and obtain

x_5	1	−1/2	−27/4	0	0	1/4
μ_1	0	1/8	−9/16	0	0	3/16
λ_3	0	0	1	0	0	1
μ_0	0	−1/8	9/16	1	0	13/16
x_0	0	2	27/2	0	1	51/2

Minimize in block I, $4x_1 + 2x_2 + 27/2 - 3x_1 - 5x_2$

$$= x_1 - 3x_2 + 27/2$$

which gives 0 in vertex ③;

Minimize in block II, $4x_3 + 2x_4 - 4x_3 - x_4 = x_4$

which gives 0 in vertex ⓪ or ①

so that we have reached the final answer. In terms of the x_i we have

$x_1 = 0,\ x_2 = 9/2$ (vertex ③ of block I)
$x_3 = 4 \times 3/16 + 0 \times 13/16 = 3/4$
$x_4 = 0 \times 3/16 + 0 \times 13/16 = 0$

We have carried out this example as an illustration of the Decomposition Algorithm. As it happens, we could have obtained, in this case, the answer more quickly by using the Simplex Method, as the following three tableaux show:

	x_1	x_2	x_3	x_4	
x_5	1	2	1	1	10
x_6	2	1	2	1	6
x_7	3	1	0	0	10
x_8	1	4*	0	0	18
x_9	0	0	1	3	8
x_{10}	0	0	3	3	12
x_0	−3	−5	−4	−1	0

	x_1	x_8	x_3	x_4	
x_5	3/2	−1/2	1	1	1
x_6	7/4	−1/4	2*	1	3/2
x_7	11/4	−1/4	0	0	11/2
x_2	1/4	1/4	0	0	9/2
x_9	0	0	1	3	8
x_{10}	0	0	3	3	12
x_0	−7/4	5/4	−4	−1	$22\frac{1}{2}$

	x_1	x_8	x_6	x_4	
x_5	5/8	−1/4	−1/2	1/2	1/4
x_3	7/8	−1/8	1/2	1/2	3/4
x_7	11/4	−1/4	0	0	11/2
x_2	1/4	1/4	0	0	9/2
x_9	−7/8	1/8	−1/2	5/2	29/4
x_{10}	−21/8	3/8	3/2	3/2	39/4
	7/4	3/4	2	1	$25\frac{1}{2}$

13 Lot Manufacturing

A customer requires five items of some commodity. The manufacturer knows that he must produce more than five items to satisfy the demand, because some of the items produced will be defective and unsaleable. If less than the required number of items is perfect in any lot, then a new lot must be set up, and the cost of any new set-up is 3 and is independent of the number of items in the lot to be produced. The production of one item costs 0·5.

We are asking for the amounts to be produced in successive lots (the number of which is unknown at the start), so that the expected total cost of production of five perfect items together with the set-up costs should be minimized. It is assumed that items produced in excess of requirement are a complete loss.

Let $p(n, m)$ be the probability that out of n items produced m items will be usable. Of course, $p(n, m) = 0$ when m exceeds n.

If $f(N)$ is the expected cost of satisfying a demand for N items, when the best policy is used, then we have

$$f(1) = \min_{n \geqslant 1}\{s + nc + p(n, 0)f(1)\}$$

where s is the set-up cost and c the cost of the production of one item. For N larger than 1, we have

$$f(N) = \min_{n \geqslant N}\{s + nc + \sum_{m=0}^{N-1} p(n, m)f(N - m)\}$$

Note that $f(N)$ appears on both sides of this functional equation. This is so, because it is possible—with probability $p(n, 0)$—that all items produced are defective, and that hence again N perfect items are required to be produced. In order to avoid trivial cases, we assume that $p(n, 0)$ is not unity.

Subtracting $p(n, 0)f(N)$ on both sides of the equations and dividing then by $1 - p(n, 0)$, we obtain

$$f(1) = \min_{n \geqslant 1} (s + nc)/(- p(n, 0))$$

and

$$f(N) = \min_{n \geqslant N} \frac{s + nc + \sum_{m=1}^{N-1} p(n, m)f(N - m)}{1 - p(n, 0)} \quad \text{for } N \geqslant 2$$

We start with $f(1)$ and compute, successively, $f(2)$, $f(3)$, etc.

We assume now that each item produced has a probability of 0·7 of being perfect, independently of whether any other item is satisfactory or not. Hence

$$p(n, m) = \binom{n}{m} (0{\cdot}3)^{n-m}(0{\cdot}7)^m$$

To aid our computations, the following table will be useful:

n	1	2	3	4	5	6	7	8	9
$m = 1$	0·7	0·42	0·189	0·0756	0·0284	0·0102	0·0036	0·0012	0·0004
2		0·49	0·441	0·2646	0·1323	0·0595	0·0250	0·0100	0·0039
3			0·343	0·4116	0·3087	0·1852	0·0972	0·0467	0·0210
4				0·2401	0·3602	0·3241	0·2269	0·1361	0·0735
5					0·1681	0·3025	0·3177	0·2541	0·1715
6						0·1176	0·2471	0·2965	0·2668
7							0·0824	0·1977	0·2668
8								0·0576	0·1556
9									0·0404
$1 - p(n, 0)$	0·7	0·91	0·973	0·9919	0·9976	0·9993	0·9998	1·0000	1·0000

We work out, successively, $f(1)$, $f(2)$, $f(3)$, $f(4)$, and eventually $f(5)$, as follows:

$$\begin{aligned} f(1) &= \min \{3{\cdot}5/0{\cdot}7[= 5], 4/0{\cdot}91[= 4{\cdot}4], 4{\cdot}5/0{\cdot}973[= 4{\cdot}6], \ldots\} \\ &= 4{\cdot}4 \\ &\text{for } n = 2 \end{aligned}$$

$$f(2) = \min\{(4 + 0{\cdot}42 \times 4{\cdot}4)/0{\cdot}91[= 6{\cdot}43], (4{\cdot}5 + 0{\cdot}189 \times 4{\cdot}4)/0{\cdot}973[= 5{\cdot}48],$$
$$(5 + 0{\cdot}0756 \times 4{\cdot}4)/0{\cdot}9919[= 5{\cdot}38],$$
$$(5{\cdot}5 + 0{\cdot}0284 \times 4{\cdot}4)/0{\cdot}9976[= 5{\cdot}64], \ldots\} = 5{\cdot}38$$

for $n = 4$

$$f(3) = \min\{(4{\cdot}5 + 0{\cdot}189 \times 5{\cdot}38 + 0{\cdot}441 \times 4{\cdot}4)/0{\cdot}973[= 7{\cdot}66],$$
$$(5 + 0{\cdot}0756 \times 5{\cdot}38 + 0{\cdot}2646 \times 4{\cdot}4)/0{\cdot}9919[= 6{\cdot}62],$$
$$(5{\cdot}5 + 0{\cdot}0284 \times 5{\cdot}38 + 0{\cdot}1323 \times 4{\cdot}4)/0{\cdot}9976[= 6{\cdot}25],$$
$$(6 + 0{\cdot}0102 \times 5{\cdot}38 + 0{\cdot}0595 \times 4{\cdot}4)/0{\cdot}9993[= 6{\cdot}32], \ldots\} = 6{\cdot}25$$

for $n = 5$

$$f(4) = \min\{(5 + 0{\cdot}0756 \times 6{\cdot}25 + 0{\cdot}2646 \times 5{\cdot}38 + 0{\cdot}4116 \times 4{\cdot}4)/0{\cdot}9919[= 8{\cdot}78],$$
$$(5{\cdot}5 + 0{\cdot}0284 \times 6{\cdot}25 + 0{\cdot}1323 \times 5{\cdot}38 + 0{\cdot}3087 \times 4{\cdot}4)/0{\cdot}9976[= 7{\cdot}77],$$
$$(6 + 0{\cdot}0102 \times 6{\cdot}25 + 0{\cdot}0595 \times 5{\cdot}38 + 0{\cdot}1852 \times 4{\cdot}4)/0{\cdot}9993[= 7{\cdot}20],$$
$$(6{\cdot}5 + 0{\cdot}0036 \times 6{\cdot}25 + 0{\cdot}0250 \times 5{\cdot}38 + 0{\cdot}0972 \times 4{\cdot}4)/0{\cdot}9998[= 7{\cdot}09],$$
$$7 + 0{\cdot}0012 \times 6{\cdot}25 + 0{\cdot}0100 \times 5{\cdot}38 + 0{\cdot}0467 \times 4{\cdot}4, \ldots\} = 7{\cdot}09$$

for $n = 7$

$$f(5) = \min\{\ldots, (6{\cdot}5 + 0{\cdot}0036 \times 7{\cdot}09 + 0{\cdot}0250 \times 6{\cdot}25 + 0{\cdot}0972 \times 5{\cdot}38 + 0{\cdot}2269 \times 4{\cdot}4)/0{\cdot}9998[= 8{\cdot}20],$$
$$7 + 0{\cdot}0012 \times 7{\cdot}09 + 0{\cdot}0100 \times 6{\cdot}25 + 0{\cdot}0467 \times 5{\cdot}38 + 0{\cdot}1361 \times 4{\cdot}4[= 7{\cdot}92],$$
$$7{\cdot}5 + 0{\cdot}0004 \times 7{\cdot}09 + 0{\cdot}0039 \times 6{\cdot}25 + 0{\cdot}0210 \times 5{\cdot}38 + 0{\cdot}0735 \times 4{\cdot}4[= 7{\cdot}96], \ldots\} = 7{\cdot}92$$

for $n = 8$

It follows that eight items should be produced to begin with, and further lots set up, if necessary, dependent on the results of previous productions. The complete scheme can be described in the following tree (where we ignore those results which have a probability of less than 0·001):

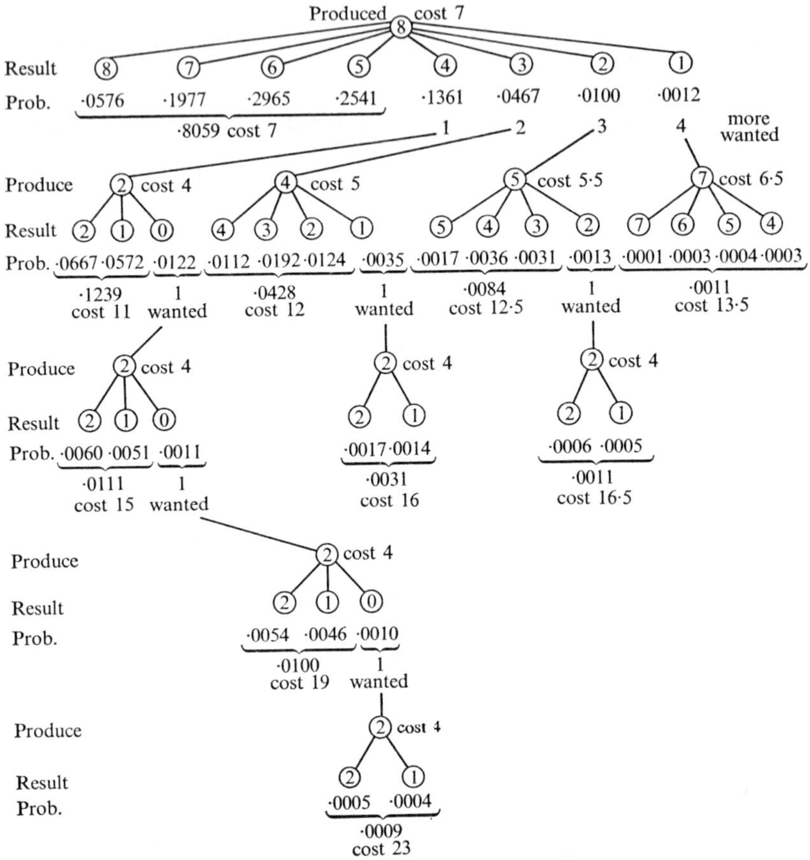

At the end of each branch of the tree, we have listed the probability of getting there and the total cost incurred up to there. The total of the products of these probabilities and cost equals $f(5)$.

This problem has been treated in [21] and in [38].

14 Selection of Brands

A firm produces five brands of cakes and makes use of three types of raw material in limited supply. The following table indicates the amounts at which the raw materials are used for the various brands, and the available supply:

	Brand					
Material	1	2	3	4	5	Available
1	7·1	6·2	5·2	6·5	7·4	20·0
2	5·7	6·3	6·8	5·0	6·3	20·0
3	2·2	2·5	3·0	3·5	1·3	25·0

The contribution to income depends on the amounts produced. In particular, if the amounts produced of the five brands are

$$x_1 \qquad x_2 \qquad x_3 \qquad x_4 \qquad x_5$$

then the contribution to income per unit is, respectively

$$12{\cdot}1 - x_1 \quad 8{\cdot}4 - 2x_2 \quad 11{\cdot}2 - 2x_3 \quad 7{\cdot}7 - x_4 \quad 10{\cdot}4 - x_5$$

provided these amounts are not negative.

It is required to make such a selection of the amounts of the brands produced that the total income is maximized.

We shall solve the problem first by the method of E. M. L. Beale,* and then by the Dantzig–van de Panne method.** The constraints are

$$7{\cdot}1x_1 + 6{\cdot}2x_2 + 5{\cdot}2x_3 + 6{\cdot}5x_4 + 7{\cdot}4x_5 + x_6 = 20{\cdot}0$$
$$5{\cdot}7x_1 + 6{\cdot}3x_2 + 6{\cdot}8x_3 + 5{\cdot}0x_4 + 6{\cdot}3x_5 + x_7 = 20{\cdot}0$$
$$2{\cdot}2x_1 + 2{\cdot}5x_2 + 3{\cdot}0x_3 + 3{\cdot}5x_4 + 1{\cdot}3x_5 + x_8 = 25{\cdot}0$$

* See [4]. ** See [36], and also [12], Chapter 24, Section 4.

and the objective function, in its symmetric form, is

$$\begin{aligned} B = {} & 6{\cdot}05x_1 + 4{\cdot}2x_2 + 5{\cdot}6x_3 + 3{\cdot}85x_4 + 5{\cdot}2x_5 \\ & + (6{\cdot}05 - x_1)x_1 \\ & + (4{\cdot}2 - 2x_2)x_2 \\ & + (5{\cdot}6 - 2x_3)x_3 \\ & + (3{\cdot}85 - x_4)x_4 \\ & + (5{\cdot}2 - x_5)x_5 \end{aligned}$$

We can take as our first basic variables the slack variables x_6, x_7, and x_8. Because at this stage all the partial derivatives of B with respect to the non-basic variables are positive (they are twice the constants in the first row of the expression for B, thus $\partial B/\partial x_1 = 12{\cdot}1$, etc.), it is useful to increase any variable, and we choose x_1. The increase of x_1 should be stopped when $x_1 = 2{\cdot}82$ (because afterwards x_6 would become negative, by virtue of the first constraint), or when it is 6·05, when the partial derivative becomes zero (as we see from the bracket expression multiplied by x_1). The first mentioned value is smaller, and therefore we exchange x_1 for x_6. The exchange is exhibited in the two tableaux:

		x_1	x_2	x_3	x_4	x_5
x_6	20·0	7·1*	6·2	5·2	6·5	7·4
x_7	20·0	5·7	6·3	6·8	5·0	6·3
x_8	25·0	2·2	2·5	3·0	3·5	1·3
B	0	−12·1	−8·4	−11·2	−7·7	−10·4

		x_6	x_2	x_3	x_4	x_5
x_1	2·82	0·14	0·87	0·73	0·92	1·04
x_7	3·93	−0·80	1·34	2·64	−0·24	0·37
x_8	18·80	−0·31	0·59	1·39	1·48	−0·99
B	34·12	1·70	2·13	−2·37	3·43	2·18

(At this stage, ignore the B-rows.)

To obtain the new form of the quadratic objective function it is convenient to proceed as follows:

First replace x_1 by $2{\cdot}82 - 0{\cdot}14x_6 - 0{\cdot}87x - 0{\cdot}73x_3 - 0{\cdot}92x_4 - 1{\cdot}04x_5$ only in the linear term and inside the brackets. (As a matter of fact, it appears in only one of these at this stage.) This gives

$$
\begin{aligned}
B = {} & 17{\cdot}04 - 1{\cdot}08x_2 + 1{\cdot}17x_3 - 1{\cdot}69x_4 - 1{\cdot}11x_5 - 0{\cdot}85x_6 \\
& + (\ 3{\cdot}23 + 0{\cdot}87x_2 + 0{\cdot}73x_3 + 0{\cdot}92x_4 + 1{\cdot}04x_5 + 0{\cdot}14x_6)x_1 \\
& + (\ 4{\cdot}2 - 2\ x_2 \qquad\qquad\qquad\qquad\qquad\qquad)x_2 \\
& + (\ 5{\cdot}6 \qquad - 2\ x_3 \qquad\qquad\qquad\qquad\qquad)x_3 \\
& + (\ 3{\cdot}85 \qquad\qquad\qquad - \ x_4 \qquad\qquad\qquad)x_4 \\
& + (\ 5{\cdot}2 \qquad\qquad\qquad\qquad\qquad - \ x_5 \qquad)x_5
\end{aligned}
$$

Now imagine the matrix transposed, thus:

$$
\begin{aligned}
& 17{\cdot}04 + 3{\cdot}23x_1 + 4{\cdot}2x_2 + 5{\cdot}6x_3 + 3{\cdot}85x_4 + 5{\cdot}2x_5 \\
& + (-1{\cdot}08 + 0{\cdot}87x_1 - 2\ x_2 \qquad\qquad\qquad\qquad)x_2 \\
& \cdot \quad \cdot \quad \cdot \quad \cdot \quad \cdot \quad \cdot \quad \cdot \quad \cdot \quad \cdot \\
& + (-0{\cdot}85 + 0{\cdot}14x_1 \qquad\qquad\qquad\qquad\qquad)x_6
\end{aligned}
$$

and substitute for x_1 in the linear term and in any of the brackets it might appear in (though it does not do so here at this stage). This gives

$$
\begin{aligned}
B = {} & 26{\cdot}14 + 1{\cdot}38x_2 + 3{\cdot}24x_3 + 0{\cdot}90x_4 + 1{\cdot}82x_5 - 0{\cdot}45x_6 \\
& + (\ 1{\cdot}38 - 2{\cdot}76x_2 - 0{\cdot}64x_3 - 0{\cdot}80x_4 - 0{\cdot}90x_5 - 0{\cdot}12x_6)x_2 \\
& + (\ 3{\cdot}24 - 0{\cdot}64x_2 - 2{\cdot}53x_3 - 0{\cdot}67x_4 - 0{\cdot}76x_5 - 0{\cdot}10x_6)x_3 \\
& + (\ 0{\cdot}90 - 0{\cdot}80x_2 - 0{\cdot}67x_3 - 1{\cdot}84x_4 - 0{\cdot}96x_5 - 0{\cdot}13x_6)x_4 \\
& + (\ 1{\cdot}82 - 0{\cdot}90x_2 - 0{\cdot}76x_3 - 0{\cdot}96x_4 - 2{\cdot}08x_5 - 0{\cdot}15x_6)x_5 \\
& + (-0{\cdot}45 - 0{\cdot}12x_2 - 0{\cdot}10x_3 - 0{\cdot}13x_4 - 0{\cdot}15x_5 - 0{\cdot}02x_6)x_6
\end{aligned}
$$

It is now appropriate to increase x_3. We have once more to consider two possible barriers to the increase. The constraints show that we must not increase x_3 beyond 1·49. But if we leave x_2, x_4, x_5, and x_6 equal to zero, then the coefficient of x_3 in the last expression for B, viz. $3{\cdot}24 - 2{\cdot}53x_3$, shows that it is not desirable to increase x_3 beyond $3{\cdot}24/2{\cdot}53 = 1{\cdot}281$. This is smaller than 1·49, so we introduce a new variable (without any sign restriction):

$$u_1 = 3{\cdot}24 - 0{\cdot}64x_2 - 2{\cdot}53x_3 - 0{\cdot}67x_4 - 0{\cdot}76x_5 - 0{\cdot}10x_6$$

and exchange x_3 and u_1, making $u_1 = 0$ in the process. In other words, we have

$$x_3 = 1{\cdot}281 - 0{\cdot}253x_2 - 0{\cdot}265x_4 - 0{\cdot}300x_5 - 0{\cdot}040x_6 - 0{\cdot}395u_1$$

and substitute this in the constraints as well as in the objective function, so that we have

$$
\begin{aligned}
x_1 &= 1{\cdot}89 - 0{\cdot}69x_2 - 0{\cdot}73x_4 - 0{\cdot}82x_5 - 0{\cdot}11x_6 + 0{\cdot}29u_1 \\
x_7 &= 0{\cdot}56 - 0{\cdot}65x_2 - 0{\cdot}93x_4 + 0{\cdot}42x_5 + 0{\cdot}90x_6 + 1{\cdot}04u_1 \\
x_8 &= 17{\cdot}02 - 0{\cdot}24x_2 - 1{\cdot}12x_4 + 1{\cdot}41x_5 + 0{\cdot}36x_6 + 0{\cdot}55u_1
\end{aligned}
$$

$$
\begin{aligned}
B = {} & 30{\cdot}29 + 0{\cdot}57x_2 + 0{\cdot}03x_4 + 0{\cdot}87x_5 - 0{\cdot}58x_6 \\
& + (\ 0{\cdot}57 - 2{\cdot}60x_2 - 0{\cdot}63x_4 - 0{\cdot}71x_5 - 0{\cdot}09x_6)x_2 \\
& + (\ 0{\cdot}03 - 0{\cdot}63x_2 - 1{\cdot}66x_4 - 0{\cdot}76x_5 - 0{\cdot}10x_6)x_4 \\
& + (\ 0{\cdot}87 - 0{\cdot}71x_2 - 0{\cdot}76x_4 - 1{\cdot}85x_5 - 0{\cdot}12x_6)x_5 \\
& + (-0{\cdot}58 - 0{\cdot}09x_2 - 0{\cdot}10x_4 - 0{\cdot}12x_5 - 0{\cdot}02x_6)u_1 \\
& + (\qquad\qquad\qquad - 0{\cdot}40u_1 \qquad\qquad\qquad)u_1
\end{aligned}
$$

It can be shown that whenever a new variable u_i is introduced, it will at first appear in this solitary state.

Now it is the turn of x_5 to be increased. The first constraint would allow us to increase it to 1·89/0·82, while the other constraints do not affect the increase at all. However, $0{\cdot}87 - 1{\cdot}85x_5$ is zero when $x_5 = 0{\cdot}469$, and this is clearly smaller. Hence we introduce

$$u_2 = 0{\cdot}87 - 0{\cdot}71x_2 - 0{\cdot}76x_4 - 1{\cdot}85x_5 - 0{\cdot}12x_6$$

i.e.

$$x_5 = 0{\cdot}469 - 0{\cdot}384x_2 - 0{\cdot}411x_4 - 0{\cdot}065x_6 - 0{\cdot}541u_2$$

and after substitutions

$$\begin{aligned}
x_1 &= 1{\cdot}51 - 0{\cdot}37x_2 - 0{\cdot}40x_4 - 0{\cdot}06x_6 + 0{\cdot}29u_1 + 0{\cdot}44u_2\\
x_3 &= 1{\cdot}14 - 0{\cdot}13x_2 - 0{\cdot}14x_4 - 0{\cdot}02x_6 - 0{\cdot}04u_1 + 0{\cdot}16u_2\\
x_7 &= 0{\cdot}76 - 0{\cdot}84x_2 - 0{\cdot}76x_4 + 0{\cdot}87x_6 + 1{\cdot}04u_1 - 0{\cdot}23u_2\\
x_8 &= 17{\cdot}68 - 0{\cdot}78x_2 - 1{\cdot}69x_4 + 0{\cdot}27x_6 + 0{\cdot}55u_1 - 0{\cdot}76u_2
\end{aligned}$$

$$\begin{aligned}
B = {} & 30{\cdot}68 + 0{\cdot}24x_2 - 0{\cdot}33x_4 - 0{\cdot}66x_6\\
& + (\;0{\cdot}24 - 2{\cdot}33x_2 - 0{\cdot}34x_4 - 0{\cdot}04x_6)x_2\\
& + (-0{\cdot}33 - 0{\cdot}34x_2 - 1{\cdot}35x_4 - 0{\cdot}05x_6)x_4\\
& + (-0{\cdot}66 - 0{\cdot}04x_2 - 0{\cdot}05x_4 - 0{\cdot}01x_6)x_6\\
& + (\qquad - 0{\cdot}40u_1 \qquad)u_1\\
& + (\qquad\qquad - 0{\cdot}54u_2 \qquad)u_2.
\end{aligned}$$

Finally, a new variable $u_3 = 0{\cdot}24 - 2{\cdot}33x_2 - 0{\cdot}34x_4 - 0{\cdot}04x_6$ is called for, because $u_3 = 0$ when $x_2 = 0{\cdot}103$, while the constraints would only restrict x_2 to the smallest of 0·469/0·384, 1·14/0·13, 1·51/0·37, 0·76/0·84, 17·68/0·78. Thus we obtain

$$\begin{aligned}
x_1 &= 1{\cdot}47 - 0{\cdot}33x_4 - 0{\cdot}05x_6 + 0{\cdot}29u_1 + 0{\cdot}44u_2 + 0{\cdot}16u_3\\
x_2 &= 0{\cdot}10 - 0{\cdot}15x_4 - 0{\cdot}02x_6 \qquad\qquad\qquad\qquad - 0{\cdot}43u_3\\
x_3 &= 1{\cdot}13 - 0{\cdot}12x_4 - 0{\cdot}02x_6 - 0{\cdot}40u_1 + 0{\cdot}16u_2 + 0{\cdot}06u_3\\
x_5 &= 0{\cdot}43 - 0{\cdot}35x_4 - 0{\cdot}06x_6 \qquad\quad - 0{\cdot}54u_2 + 0{\cdot}16u_3\\
x_7 &= 0{\cdot}67 - 0{\cdot}99x_4 + 0{\cdot}88x_6 + 1{\cdot}04u_1 - 0{\cdot}23u_2 + 0{\cdot}36u_3\\
x_8 &= 17{\cdot}60 - 1{\cdot}58x_4 + 0{\cdot}28x_6 + 0{\cdot}55u_1 - 0{\cdot}76u_2 + 0{\cdot}33u_3
\end{aligned}$$

$$\begin{aligned}
B = {} & 30{\cdot}70 - 0{\cdot}35x_4 - 0{\cdot}64x_6\\
& + (-0{\cdot}35 - 1{\cdot}30x_4 - 0{\cdot}04x_6)x_4\\
& + (-0{\cdot}64 - 0{\cdot}04x_4 - 0{\cdot}01x_6)x_6\\
& -0{\cdot}40u_1 - 0{\cdot}54u_2 - 0{\cdot}43u_3
\end{aligned}$$

We proceed to the solution of the problem by the Dantzig–van de Panne method.

The set-up tableau is

	x_1	x_2	x_3	x_4	x_5	y_6	y_7	y_8	
y_1	−2	0	0	0	0	−7·1	−5·7	−2·2	−12·1
y_2	0	−4	0	0	0	−6·2	−6·3	−2·5	−8·4
y_3	0	0	−4	0	0	−5·2	−6·8	−3·0	−11·2
y_4	0	0	0	−2	0	−6·5	−5·0	−3·5	− 7·7
y_5	0	0	0	0	−2	−7·4	−6·3	−1·3	−10·4
x_6	7·1*	6·2	5·2	6·5	7·4	0	0	0	20·0
x_7	5·7	6·3	6·8	5·0	6·3	0	0	0	20·0
x_8	2·2	2·5	3·0	3·5	1·3	0	0	0	25·0
F	12·1	8·4	11·2	7·7	10·4	20·0	20·0	25·0	0

and it is, as every set-up tableau, in standard form. The rules tell us to exchange x_1 for x_6 (and not for y_1, because 20·0/7·1 is smaller than 12·1/2), so that the next tableau is non-standard, thus:

	x_6	x_2	x_3	x_4	x_5	y_6	y_7	y_8	
y_1	0·28	1·74	1·46	1·84	2·08	−7·1*	−5·7	−2·2	−6·46
y_2	0	−4	0	0	0	−6·2	−6·3	−2·5	−8·4
y_3	0	0	−4	0	0	−5·2	−6·8	−3·0	−11·2
y_4	0	0	0	−2	0	−6·5	−5·0	−3·5	−7·7
y_5	0	0	0	0	−2	−7·4	−6·3	−1·3	−10·4
x_1	0·14	0·87	0·73	0·92	1·04	0	0	0	2·82
x_7	−0·80	1·34	2·64	−0·24	0·37	0	0	0	3·93
x_8	−0·31	0·59	1·39	1·48	−0·99	0	0	0	18·80
F	−1·70	−2·13	2·37	−3·43	−2·18	20·0	20·0	25·0	−34·12

We continue by exchanging y_6 and y_1, which produces again a standard tableau. Further exchanges are x_3 for y_3, x_5 for y_5, and finally x_2 for y_2, which produces the optimal tableau:

	x_6	y_2	y_3	x_4	y_5	y_1	y_7	y_8	
y_6	−0·02	−0·02	−0·02	−0·11	−0·05	−0·05	0·89	0·29	1·28
x_2	0·02	−0·22	0·03	0·15	0·08	0·08	0·18	0·17	0·10
x_3	0·02	0·03	−0·22	0·12	0·07	0·06	0·53	0·36	1·13
y_4	−0·11	−0·15	−0·12	−2·62	−0·35	−0·35	0·88	−1·57	0·73
x_5	0·05	0·08	0·07	0·35	−0·30	0·19	−0·18	−0·45	0·43
x_1	0·05	0·08	0·06	0·35	0·19	−0·31	0·36	0·05	1·47
x_7	−0·89	0·18	0·53	−0·88	−0·18	−0·36	−1·57	−1·02	0·67
x_8	−0·29	0·17	0·36	1·57	−0·45	0·05	−1·02	−1·05	17·60
	−1·28	0·10	1·13	−0·73	0·43	1·47	0·67	17·60	−61·41

It is of some interest to compare this result with that which we obtain if the contribution to income per unit is a constant, say respectively

12·1	8·4	11·2	7·7	10·4

Because in the quadratic case the rate of contributions to income decrease with increasing x_i, it is to be expected that the maximal income will be smaller in the quadratic case than in the linear one. The latter starts off with the two tableaux which we have already exhibited and then continues:

		x_6	x_2	x_7	x_4	x_5
x_1	1·73	0·36	0·50	−0·28	0·99	0·94
x_3	1·49	−0·30	0·51	0·38	−0·09	0·14
x_8	16·72	0·11	−0·12	−0·53	1·61	−1·18
B	37·62	0·99	3·34	0·89	3·22	2·51

Thus the largest obtainable income is 37·62 and to this end only the first and third varieties should be manufactured, at levels of 1·73 and 1·49. Of course, the policy of the firm might exclude such a restrictive decision.

Let us compare the solutions to the two cases:

	Linear	*Quadratic*
	objective function	
x_1	1·73	1·47
x_2	0	0·10
x_3	1·49	1·13
x_4	0	0
x_5	0	0·43
x_6	0	0
x_7	0	0·67
x_8	16·72	17·60
B	37·62	30·70

It will be noticed that in the quadratic case the optimum is not obtained at a basic solution of the constraints, as it is with a linear objective function.

When we compare the successive tableaux with the steps in Beale's method, then we see that the two methods are closely related. This parallelism was studied in [37].

15 Mixed Transport

A merchant has three warehouses and stores in them, respectively, 12·4, 8·7, and 15·4 tons of grain. He has three customers who require 12·1, 7·4, and 17·0 tons. Between each customer and each warehouse there is a choice of two methods of transport with differing costs; method 1 can carry goods only in multiples of one ton, while no such restriction applies to method 2. It is required to determine the cheapest way of supplying the goods, at the following costs per ton:

Method 1				Method 2		
To customer				To customer		
1	2	3	Warehouse	1	2	3
5	6	3	1	6	4	5
3	1	7	2	2	4	6
8	8	6	3	6	7	7

We start off by ignoring the requirement of integrality in the first method and solve the problem assuming that we can use the cheaper method of the two for any warehouse–customer link. This leads to the Transportation Problem with the following cost table (the marginal totals are also indicated):

	12·1	7·4	17·0
12·4	5(i)	4	3(i)
8·7	2	1(i)	6
15·4	6	7	6(i)

where (i) means that the cost is that of method 1.

The solution, verified by the fictitious costs indicated, is

	12·1	7·4	17·0	f.c.
12·4			12·4	0
8·7	1·3	7·4		2
15·4	10·8		4·6	3
f.c.	0	−1	3	

However, this can not be the final solution because all these values are fractional, even those which would be transported by method 1.

If the amounts transported by method 1 are denoted by x_{ij}, and those transported by method 2 are denoted by y_{ij}, then our original problem can be written

$$x_{11} + x_{12} + x_{13} + y_{11} + y_{12} + y_{13} = 12{\cdot}4$$
$$x_{21} + x_{22} + x_{23} + y_{21} + y_{22} + y_{23} = 8{\cdot}7$$
$$x_{31} + x_{32} + x_{33} + y_{31} + y_{32} + y_{33} = 17{\cdot}0$$

This is a simple transportation problem, but for the variables $x_{ij} + y_{ij}$ rather than for x_{ij}. We have, in fact, solved the problem for

$$x_{11}, y_{12}, x_{13}$$
$$y_{21}, x_{22}, y_{33}$$
$$y_{31}, y_{32}, x_{33}$$

having ignored, for the time being, the remaining variables.

To continue, we form our solution into a Simplex Tableau. The basic variables are x_{13}, y_{21}, x_{22}, y_{31}, and x_{33} and the non-basic ones are x_{11}, y_{12}, y_{23}, y_{32}. To construct the tableau, we express the basic variables in terms of the non-basic ones. This is simple, because the constraints of a transportation problem are triangular in the basic variables. We obtain:

$$x_{13} = 12{\cdot}4 - x_{11} - y_{12}$$
$$y_{21} = 8{\cdot}7 - x_{22} - y_{23} = 1{\cdot}3 + y_{12} + y_{32} - y_{23}$$
$$x_{22} = 7{\cdot}4 - y_{12} - y_{32}$$
$$y_{31} = 12{\cdot}1 - x_{11} - y_{21} = 10{\cdot}8 - x_{11} - y_{12} - y_{32} + y_{23}$$
$$x_{33} = 15{\cdot}4 - y_{31} - y_{32} = 4{\cdot}6 + x_{11} + y_{12} - y_{23}$$

Hence the Simplex Tableau, together with the row of the objective function, is as follows:

		5	4	6	7	
		x_{11}	y_{12}	y_{23}	y_{32}	
3	x_{13}	1	1			12·4
2	y_{21}		−1	1	−1	1·3
1	x_{22}		1		1	7·4
6	y_{31}	1	1	−1	1	10·8
6	x_{33}	−1	−1	1		4·6
		−2	−2	−4	−2	139·6

From this, we construct the complete tableau, for all the x and y variables. In the row of a x_{ij} (or a y_{ij}), we enter 1 in the column of the corresponding y_{ij} (or x_{ij}). Moreover, the above entries are repeated in the corresponding columns. The row of the objective function is found again by subtracting the column cost from the entries in the column multiplied by the appropriate row cost. We obtain:

		5	6	3	7	8	8	6	4	5	4	6	7	7	
		x_{11}	x_{12}	x_{21}	x_{23}	x_{31}	x_{32}	y_{11}	y_{12}	y_{13}	y_{22}	y_{23}	y_{32}	y_{33}	
3	x_{13}	1	1					1	1	1					12·4
1	x_{22}		1				1		1		1		1		7·4
6	x_{33}	−1	1		1			−1	−1			1		1	4·6
2	y_{21}		−1	1	1		−1		−1			1	−1		1·3
6	y_{31}	1	1		−1	1	1	1	1			−1	1		10·8
	C	−2	−4	−1	−5	−2	−3	−3	−2	−2	−3	−4	−2	−1	139·6

The first three variables ought to be integers. We observe the first row and add the new variables s_1, with its row:

s_1	−1	−1					−1	−1	−1*				−0·4

This is transformed into

	x_{11}	x_{12}	x_{21}	x_{23}	x_{31}	x_{32}	x_{11}	y_{12}	s_1	y_{22}	y_{23}	y_{32}	y_{33}	
x_{13}									1					12·0
x_{22}		1				1		1		1		1		7·4
x_{33}	−1	−1		1			−1	−1			1		1	4·6
y_{21}		−1	1	1		−1		−1			1	−1		1·3
y_{31}	1	1		−1	1	1	1	1			−1	1		10·8
y_{13}	1	1					1	1	−1					0·4
C	0	−2	−1	−5	−2	−3	−1	0	−2	−3	−4	−2	−1	140·4

and finally, with the additional row:

s_2		−1				−1		−1*		−1		−1		−0·4

into

	x_{11}	x_{12}	x_{21}	x_{23}	x_{31}	x_{32}	y_{11}	s_2	s_1	y_{22}	y_{23}	y_{32}	y_{33}	
x_{13}									1					12·0
x_{22}								1						7·0
x_{33}	−1			1		1	−1	−1		1	1	1	1	5·0
y_{21}			1	1				−1		1	1			1·7
y_{31}	1			−1	1		1	1		−1	−1			10·4
y_{13}	1					−1	1	1	−1	−1		−1		0
y_{12}		1				1		−1		1		1		0·4
C	0	−2	−1	−5	−2	−3	−1	0	−2	−3	−4	−2	−1	140·4

We interpret this result to mean that the cheapest way of transporting the goods costs 140·4, and is done in the following manner:

	Method 1			Totals of both methods combined			Method 2		
	0	7	17	Total		Total	12·1	0·4	0
			12	12	12·4	0·4		0·4	
		7		7	8·7	1·7	1·7		
			5	5	15·4	10·4	10·4		
Cost				73	140·4	67·4			

The tableau suggests alternative solutions with y_{13} non-basic, but since it has already the value 0, the alternatives mean, when interpreted, the same scheme of transport.

The algorithm for mixed integer programming used above is explained in [23].

16 Flow Through a Network

Consider the following network, where the links have the stated capacities in the directions indicated by the arrows. The source S and the terminal T have unlimited capacities. It is required to send, from S to T, the largest possible flow which the network can support.

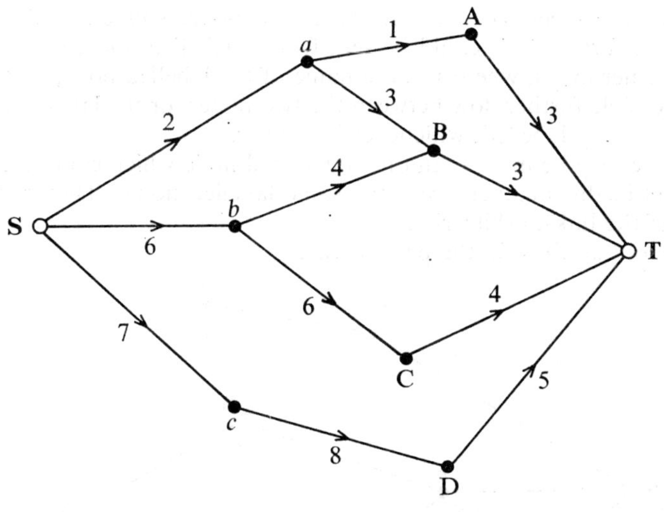

We start by choosing a flow which is obviously possible. (If we can not think of anything better, a zero flow through all links is clearly possible.) In the present case, we might start off with the network shown at the top of the next page.

We must now find out whether this is the largest possible flow. For this purpose we apply a labelling technique.

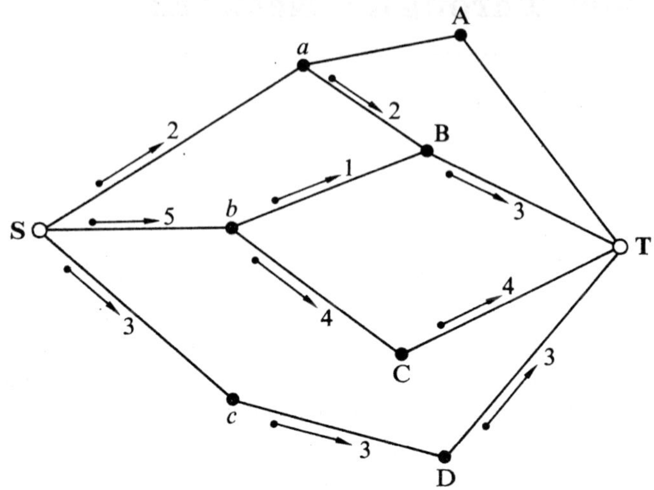

First Labelling Rule

Attach the label ∞ to S. If there is a node, still unlabelled, to which a further flow could be sent from a labelled node, then label the former (x, y), where y is the name of the labelled node, and x is the possible further flow between the two nodes, or the label y of the node already labelled, whichever is smaller.

There may be more than one unlabelled node which could receive a label in this manner, from the same labelled node. Then choose any of the links, arbitrarily.

We obtain thus, in the present case:

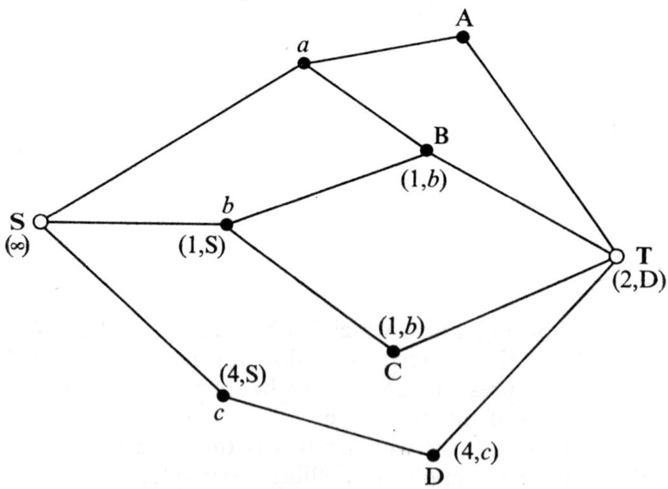

We have now labelled T. This indicates that a larger flow could be led into T, and the labels indicate that 2 more could be sent to it from D and that D could be fed 4 more from *c*, but since this could not be led any further, we let only 2 (from the label of T) flow from *c* into D, and 2 from S into *c*. This produces a flow which is larger than before by 2. We give it in the following diagram, where the new labels are also indicated:

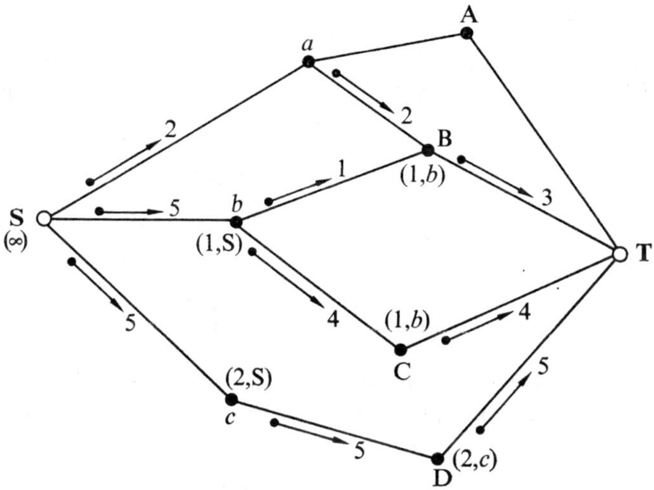

Now it is not possible to label T. However, this does not necessarily mean that the largest flow has already been reached. It is conceivable that by starting from another initial flow, we could have done better. We must investigate whether a re-arrangement of flows would help. To do this, we apply the second labelling rule.

Second Labelling Rule

If there is a positive flow from an unlabelled node into a labelled one, then label the former (z, y), where y is the name of the latter (the labelled) node, and z is the flow between the two nodes, or the label x of the labelled node, whichever is smaller.

This means that we reduce the flow by z, and we can do this because the labelling has shown that z can be provided for in another way.

In our example, B is labelled, but *a* is not. Since there is a flow of 2 from *a* to B, and the label of B is $(1, b)$, *a* can be labelled $(1, \text{B})$.

From now on we apply either of the two rules to see if we can reach T. In the present case A can be labelled (1, *a*), and then T (1, A), thus:

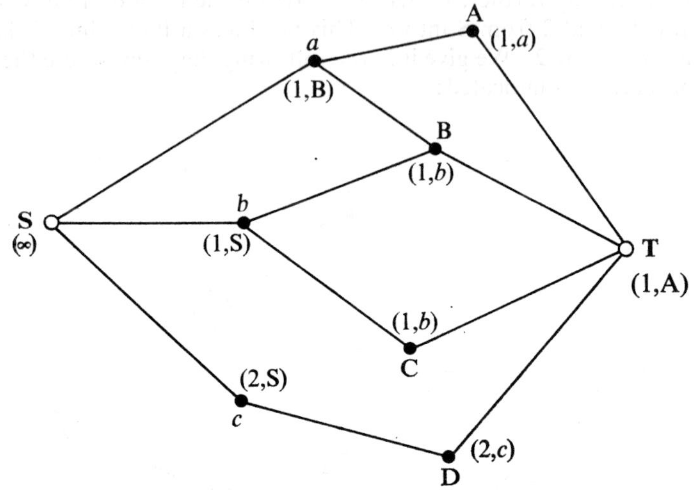

Tracing the labels backwards from T, and remembering where to withdraw a flow from a link and how to replace it, we obtain:

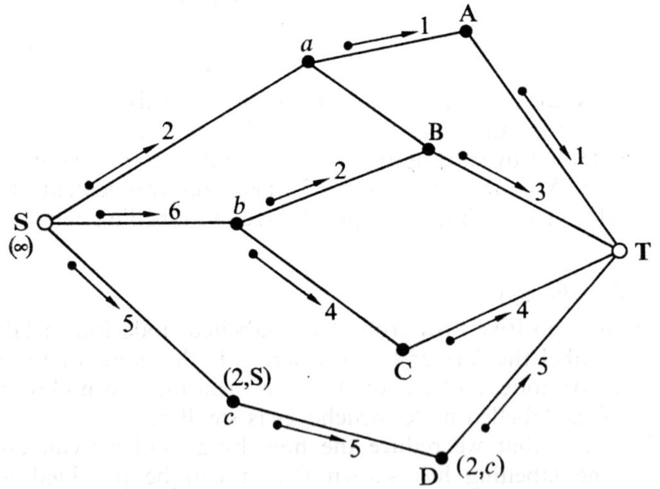

We re-start labelling, and the new labels are also given in the diagram above.

It is now impossible to reach T by either of the labelling rules. We proceed to show that it follows that the maximum flow (13 in this case) has been obtained.

For the proof, divide the set of all nodes into two subsets, that of the labelled nodes (S, *c*, D) and that of the unlabelled ones (*a*, *b*, A, B, C, T). Consider the links which connect a node of the first to a node of the second set. They are said to form a cut. (In this case they are S*a*, S*b*, and DT.) The flow in each link reaches the full capacity of that link—since otherwise we could label the end-points. No flow passes from any node of the second set to a node of the first set, because otherwise we could label further nodes. Hence the total flow equals the sum of the capacities of the links in the cut. If we omitted these links, no flow from S to T would be possible, and because those links are all used to full capacity, no larger flow is possible. It follows that the maximal flow has been found.

The labelling method is due to the authors of [19].

In the problem we have just treated there was one source and one terminal. Now we want to consider the so-called "multi-commodity" case, where we have source-and-destination pairs and flows to be sent from each source to each destination with which it forms a pair.

For instance, take the following network, with capacities of the links as indicated.

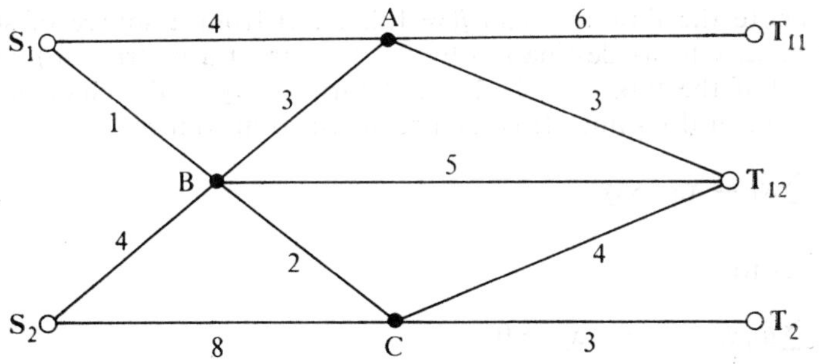

A flow is to be sent from S_1 to T_{11}, from S_1 to T_{12}, and also from S_2 to T_2. We want to find the largest possible total flow from the sources to their respective destinations. This total flow is restricted by the fact that the sum of flows through any link—in either direction—must not exceed the capacity of that link.

The name multi-commodity flow is due to the idea that each pair of source and destination, (S_1T_{11}), (S_1T_{12}), and (S_2T_2), refers to different commodities, and we shall occasionally use this term.

Clearly, if each source were related to each destination, then the problem could be solved by the labelling method of the first part of this chapter, after connecting all sources (destinations) to a master source (master destination) by links of unrestricted capacity. Thus, if in the above network it were also required to send some "commodity" from S_1 to T_2, and some from S_2 to T_{11}, and also to T_{12}, then the total flow would amount to 17.

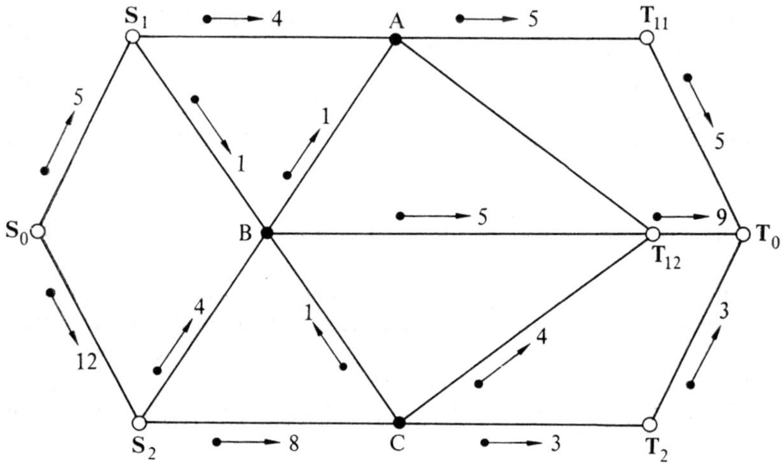

Denote the flow in chain $j(= 1,2, \ldots)$ from a source of some commodity to its destination by x_j. Construct a matrix (a_{ij}) where $a_{ij} = 1$ if the link i $(= 1, 2, \ldots)$ forms part of the chain j, and $a_{ij} = 0$ if it does not. It is then required to maximize

$$\sum_j x_j = x_0 \quad \text{say}$$

subject to

$$\sum_j a_{ij}x_j \leqslant c_i \qquad x_j \geqslant 0$$

where c_i is the capacity of link i.

This is a problem of linear programming.

However, there may be a great number of chains in the network, and we intend to avoid enumerating them all and to construct the matrix (a_{ij}) explicitly, to begin with. We use, therefore, again an approach that has served us well in the chapter on trim loss reduction.

We start with the basic feasible solution containing the slack variables, and the identity matrix, as follows:

S_1A	1												4	1*
S_1B		1											1	0
S_2B			1										4	0
S_2C				1									8	0
BA					1								3	0
BC						1							2	0
AT_{11}							1						6	1
AT_{12}								1					3	0
BT_{12}									1				5	0
CT_{12}										1			4	0
CT_2											1		3	0
X_0												1	0	−1

where S_1A, S_1B, etc., mean the slack variables whose values equal the capacities of the links of the same description. (The last column will be discussed presently.)

The columns not yet included in the tableau are those of the chains, which we describe also by the nodes through which they pass.

At this first stage any flow through some chain will increase the total which is now zero. We choose, arbitrarily, the chain S_1AT_{11}, whose column has 1 in the rows of S_1A and at AT_{11}, −1 in the row of the objective function (the last row), and everywhere else zero. This is the additional right-hand column above. The pivot is the entry 1 in the first row, marked with an asterisk. The next Simplex tableau differs from the last in the first column, and the last column. They are now as follows:

			S_1BAT_{11}
S_1AT_{11}	1	4	0
S_1B	0	1	1*
S_2B	0	4	0
S_2C	0	8	0
BA	0	3	1
BC	0	2	0
AT_{11}	−1	2	1
AT_{12}	0	3	0
BT_{12}	0	5	0
CT_{12}	0	4	0
CT_2	0	3	0
	1	4	−1

We have now to ascertain if a column of this stage, which is not exhibited, has a negative value in the last row, and we want to do this without constructing all those columns.

At the first stage all those columns which referred to chains (and were not exhibited) had -1 in the last row. At the present stage this value is found by the procedure of the Inverse Matrix Method, i.e. we multiply the original values of the column, say

$$a_{1j}, \quad a_{2j}, \quad \ldots \quad a_{11j}, \quad -1$$

by the values in the last row of the present tableau, i.e.

$$1 \quad 0 \quad \ldots 0 \quad 1$$

To know if amongst these values, for all columns, there is a negative one, we might ask which is the smallest value which we obtain in this way, and whether this smallest value is negative.

The value for a particular chain is seen to be the length of that chain, reduced by 1 if the links have the values in the last row of the tableau. We can therefore say that we are looking for the shortest chain, of any commodity, in the given network with length 1 of the link S_1A, and length 0 of the other links.

There exist many convenient procedures for finding the shortest chain in a network. In the present example we can find it by inspection.

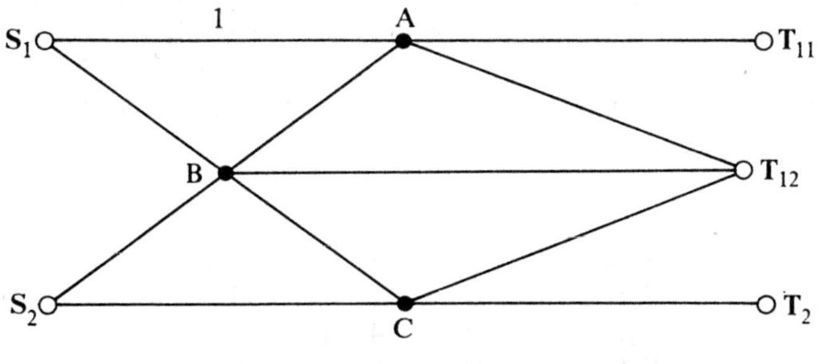

For instance, the length of the chain S_1BAT_{11} is zero, and reducing this by 1 gives a negative answer; we introduce this chain. We have written down its original column above. It must be transformed by the rule of the Inverse Matrix Method, but as it happens this leaves it unchanged in the present example. Its pivot is in the second row. The next Simplex tableau differs therefore from the original tableau, in the first two and in the last column. (The first column is the same as in the last tableau.)

We write them down

				S_2CT_2
S_1AT_{11}	1	0	4	0
S_1BAT_{11}	0	1	1	0
S_2B	0	0	4	0
S_2C	0	0	8	1
BA	0	−1	2	0
BC	0	0	2	0
AT_{11}	−1	−1	1	0
AT_{12}	0	0	3	0
BT_{12}	0	0	5	0
CT_{12}	0	0	4	0
CT_2	0	0	3	1*
	1	1	5	−1

We are now looking once more for a chain whose length is smaller than 1, if the lengths of S_1A and of S_1B are 1, and those of the other links are zero.

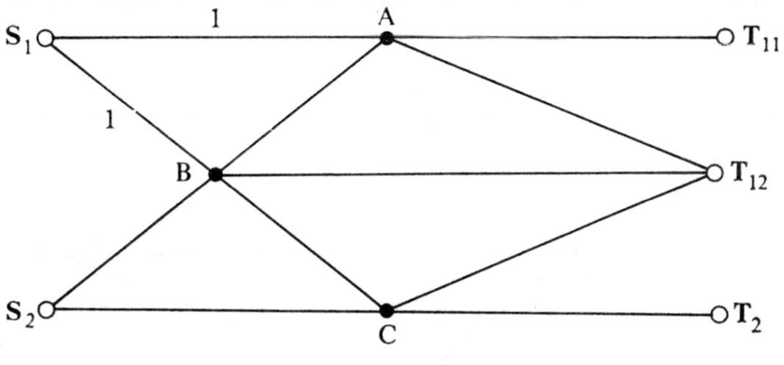

A chain of the required property is S_2CT_2, and we have added its column, which is the same in the original, and in the present, third tableau. The pivot is in the CT_2 row, and we obtain:

S_1AT_{11}	1	0	0	0	0	0	0	0	0	0	0	0	4
S_1BAT_{11}	0	1	0	0	0	0	0	0	0	0	0	0	1
S_2B	0	0	1	0	0	0	0	0	0	0	0	0	4
S_2C	0	0	0	1	0	0	0	0	0	0	—1	0	5
BA	0	−1	0	0	1	0	0	0	0	0	0	0	2
BC	0	0	0	0	0	1	0	0	0	0	0	0	2
AT_{11}	−1	−1	0	0	0	0	1	0	0	0	0	0	1
AT_{12}	0	0	0	0	0	0	0	1	0	0	0	0	3
BT_{12}	0	0	0	0	0	0	0	0	1	0	0	0	5
CT_{12}	0	0	0	0	0	0	0	0	0	1	0	0	4
S_2CT_2	0	0	0	0	0	0	0	0	0	0	1	0	3
	1	1	0	0	0	0	0	0	0	0	1	1	8

There is now no chain left with length less than 1, so that we have reached the maximal flow, 8. (There are chains with length precisely

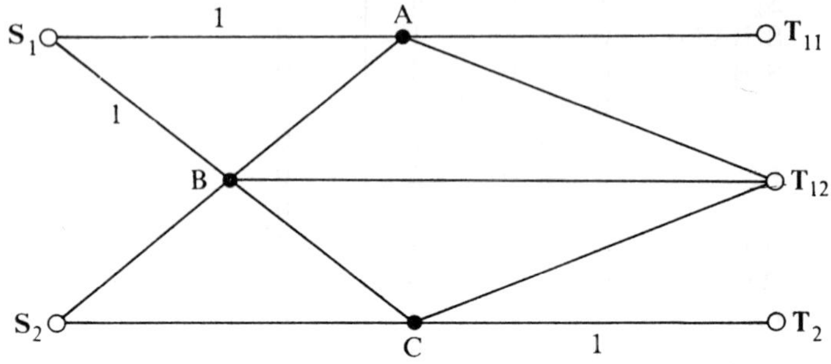

1, and hence other ways of obtaining 8, but no way of obtaining more.) The flow is made up of

4 in S_1AT_{11}, 1 in S_1BAT_{11}, and 3 in S_2CT_2.

The other entries in the last column indicate residual capacities in the links.

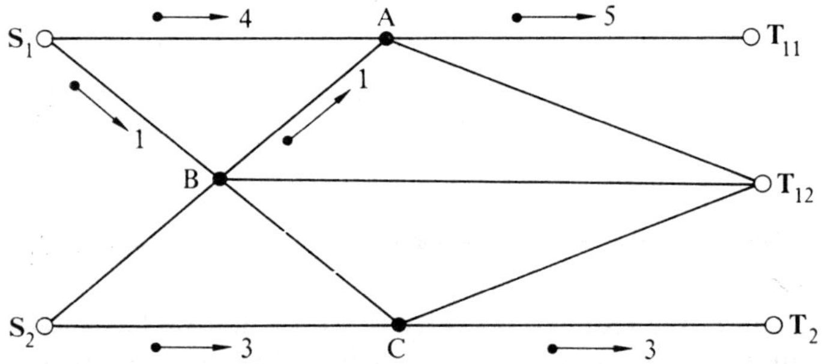

We have here followed the theory in [41].

17 Flow of Transport

We return once more to the Transportation Problem which we have solved in Chapter 1. The following table shows, in the marginal totals, availabilities and requirements, and in the body of the table the "cost" of using any of the cells indicating the routes from the place of the availabilities to the place where a facility is required:

	3	3	4	5
2	13	11	15	20
6	17	14	12	13
7	18	18	15	12

We shall now solve this problem (of finding the cheapest distribution) by a series of solutions of maximum flow problems.

We remember the constraints

$$\sum_j x_{ij} = a_i \qquad \text{and} \qquad \sum_i x_{ij} = b_j$$

and note that the solution of a problem with costs c_{ij} is the same as the solution of a problem with costs $c_{ij} - c_i - d_j$, where c_i depends only on the row and d_j only on the column of the cell (i, j). (The total cost will, of course, be different.) This is so, because the total cost to be minimized of the modified problem is

$$\sum_i \sum_j c_{ij} x_{ij} - \sum_i a_i c_i - \sum_j b_j d_j$$

and the last two terms are constants, independent of the x_{ij}, so that the complete expression is minimized by the same set of x_{ij} as the first term on its own.

We choose the c_i and d_j in such a way that all $c_{ij} - c_i - d_j \geqslant 0$, and look for the largest flow when we use those routes only for which $c_{ij} - c_i - d_j = 0$. If this largest flow equals $\sum_i a_i = \sum_j b_j$, then the solution of the original problem has been obtained, because using routes for which $c_{ij} - c_i - d_j > 0$ would be more expensive. But if the total flow, viz. $\sum_i \sum_j x_{ij}$, is smaller than the total of the marginal values, $\sum_i a_i (= \sum_j b_j)$, then we must change c_i and/or b_j in such a way that other routes can be used, while $c_{ij} - c_i - d_j \geqslant 0$ remains valid. We have every reason to be confident that appropriate values of c_i and d_j to give the final answer exist, since the fictitious costs of the final answer (see Chapter 1) satisfy precisely the conditions which c_i and d_j must satisfy.

It is convenient to choose as the first c_i the smallest c_{ij} of the i-th row, and then the d_j as the smallest $c_{ij} - c_i$ of the j-th column. In the present example we obtain thus the following cost tables:

2	0	4	9	$(c_1 = 11)$
5	2	0	1	$(c_2 = 12)$
6	6	3	0	$(c_3 = 12)$

and then

	3	3	4	5
2	0	0	4	9
6	3	2	0	1
7	4	6	3	0

$(d_1 = 2)$

Graphically, we can describe the routes to be used (corresponding to the cells with 0 entry) as follows:

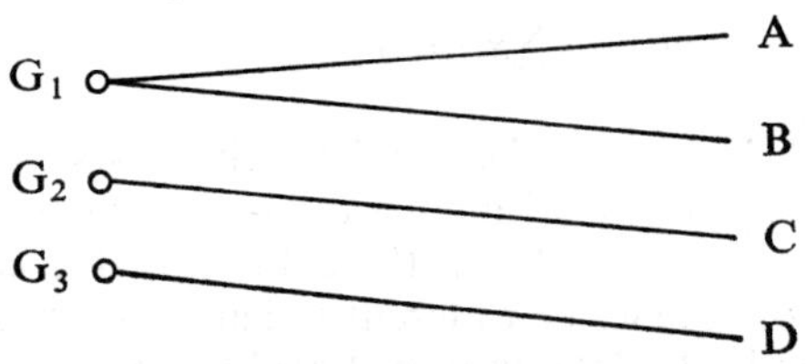

However, this graph does not give any indication of the availabilities and of the requirements. They can be taken into account by introducing a "master source" S, and a "master terminal" T, and then giving the appropriate capacities to their links, as follows:

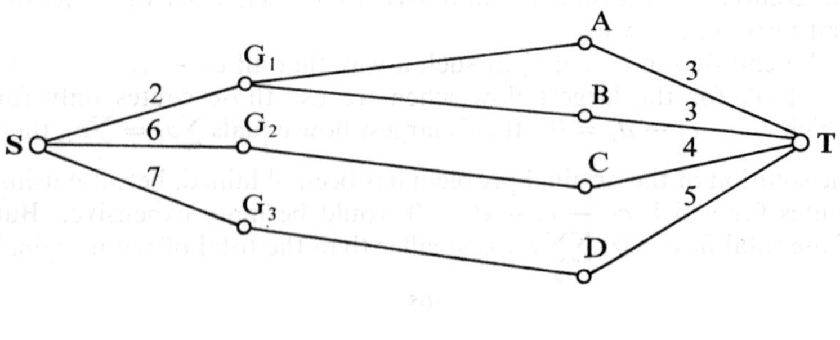

The capacities of the intermediate arcs are infinite or, if we wish, as large as $\sum a_i = \sum b_j = 15$, since no larger capacity will, in any case, be required.

It is clear from the graph, that the largest possible flow in this network is $2 + 4 + 5 = 11$. However, instead of using a graph-labelling technique, we apply an equivalent table-labelling method, because the graph describing a transportation problem is always of such a simple type that it can easily be translated into tabular presentation. Such a graph has links from a master source to certain nodes, links from these nodes to some others—but never links between nodes of the same set—and finally links from the second set of nodes to a master terminal. In the table above, zero entries correspond to the intermediate links, with infinite capacities; the marginal entries in the first vertical column correspond to the capacities of the links to nodes in the first set; and the marginal totals in the first horizontal row to the capacities of the links from the nodes of the second set. It is easy to adapt the labelling method to a tabular representation of this type.

We start with some possible flow, and its labels (writing 0 for S):

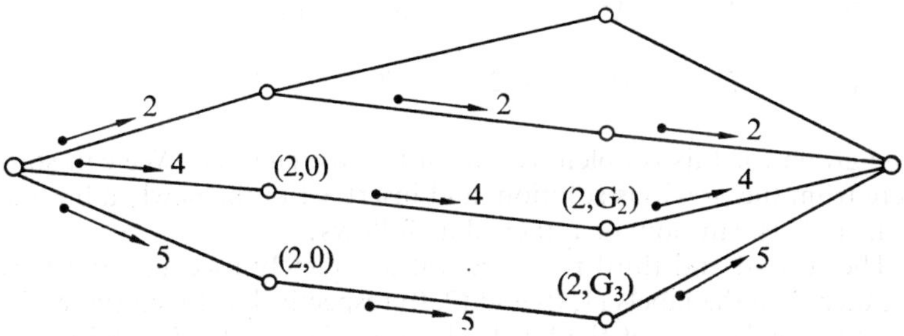

	3	3	4	5	c_i	
2	0	2	X	X	11	
6	X	X	4	X	12	(2, 0)
7	X	X	X	5	12	(2, 0)
d_j	2	0	0	0		
			(2, 2)	(2, 3)		

Admissible links of the graph corresponds to those cells where $c_{ij} - c_i - d_j = 0$.

The labels of G_2 and G_3 appear, in the table, in rows 2 and 3. They tell us that these sources have unused capacities which are still available. The column labels reflect the fact that there is an admissible link between the node corresponding to the column, and the previously labelled row. But these columns are already fully covered, which means that we can not label the terminal in the graph.

The total of all present entries is 11, which is smaller than the total to be transported (viz. 15); therefore we must change at least some of the c_i and d_j. For this purpose we consider now all cells in

labelled rows and unlabelled columns

The smallest $c_{ij} - c_i - d_j$ in these cells is 2, and therefore we

increase the c_i of labelled cells and
decrease the d_j of labelled columns

by 2. (The reason for, and the validity of, this rule are explained in more detailed descriptions. See, for instance [18].)

The new problem is then, in tabular form:

	3	3	4	5	c_i
2	0	2	*X*	*X*	11
6	*X*	1	4	*X*	14
7	*X*	*X*	*X*	5	14
d_j	2	0	−2	−2	

The graph for this problem is that of the last chapter. We can start here from the previous solution, and insert a 1 in the newly admitted cell. Labels can now be attached as follows:

The second and third rows are not completely used up, and they get therefore the labels (1, 0) and (2, 0) respectively. From these, the last three columns get the labels (1, 2), (1, 2), and (2, 3). All this is an application of the First Labelling Rule. Once more, we cannot label the terminal. However, we apply now the Second Labelling Rule. This is interpreted as meaning that if, in a column already labelled, there is a positive entry in an as yet unlabelled row, then this row is to be labelled. In the present case the second column leads to a label (1, 2) for the first row, and hence to a label (1, 1) for the first column. We have obtained:

	3	3	4	5	
2	0	2			(1, 2)
6		1	4		(1, 0)
7				5	(2, 0)
	(1, 1)	(1, 2)	(1, 2)	(2, 3)	

These labels appeared also in the last chapter, and since the first, as yet not exhausted column is now labelled, we can in the graph label the terminal, and the flow can be adjusted, as the labels indicate. As in the last chapter, we obtain

	3	3	4	5	
2	1	1			
6		2	4		
7				5	(2, 0)
				(2, 3)	

With the given graph, this is the largest flow. However, we have still not reached the answer to the original problem.

The smallest $c_{ij} - c_i - d_j$ of the cells in labelled rows and unlabelled columns is now 2, and thus we are led to

	3	3	4	5	c_i
2	1	1	*X*	*X*	11
6	*X*	2	4	*X*	14
7	2	*X*	*X*	5	16
d_j	2	0	−2	−4	

This is the final answer because there are no deficiencies left in either row or column. It agrees, of course, with the answer in Chapter 1. The c_i are larger by 11 and the d_j smaller by 11 than the fictitious costs in the final table of that chapter.

We shall refer to the method of this chapter as the "Network Method." This method was introduced in [18].

18 Personnel Assignment

There are four jobs and four candidates for them. It is possible to give rating numbers to each candidate for each job, expressing the levels of their suitabilities. It is required to allocate the candidates to jobs in such a way that the total of the four ratings, taking for each person that of the job to which he is assigned, is as high as possible.

The rating numbers are as follows:

		Jobs 1	2	3	4
Candidates	1	7	9	5	0
	2	3	6	8	7
	3	2	2	5	8
	4	2	2	5	8

We denote the candidates by $i = 1, 2, 3$, or 4, and the jobs by $j = 1, 2, 3, 4$. The rating number of candidate i in job j will be denoted by r_{ij}. Let x_{ij} equal 1, if candidate i is allocated to job j, and zero otherwise. Then the x_{ij} must be non-negative, and satisfy the following equations:

$$\sum_j x_{ij} = 1 \text{ for all } i \qquad \sum_i x_{ij} = 1 \text{ for all } j$$

It is required to find the maximum of

$$\sum_{i=1}^{4} \sum_{j=1}^{4} r_{ij}x_{ij}$$

This is clearly a problem of transportation type. It would not be advisable to try and solve it by the stepping-stone method, because the scheme is highly degenerate. Indeed, the solution will only contain four positive variables, instead of $4 + 4 - 1 = 7$. On the other hand, the network method is eminently suitable for this case.

It is more convenient to work with an objective function to be minimized, and therefore we introduce complementary rating numbers, i.e. the difference between the numbers in the above table and a suitable high number, say 10.

Complementary rating numbers:

	$j = 1$	2	3	4
$i = 1$	3	1	5	10
2	7	4	2	3
3	8	8	5	2
4	8	8	5	2

The network method works as follows (all the row and column totals are, of course, 1 and will not be indicated):

					c_i	
	1	.	+	+	1	
	+	+	1	+	2	
	+	+	+	1	2	(1, 4)
	+	+	+	.	2	(1, 0)
d_j	2	0	0	0		
				(1, 4)		

	1	.	+	+	1	(1, 1)
	+	+	1	+	2	
	.	+	+	1	6	
	.	+	+	.	6	(1, 0)
d_j	2	0	0	−4		
	(1, 4)	(1, 1)		(1, 4)		

.	1	+	+
+	+	1	+
1	+	+	.
.	+	+	1

The last table contains the answer to our problem. If we allocated candidates 1, 2, 3, and 4 to jobs 2, 3, 1, and 4, in this order, then the total of their rating numbers is 27, and no higher total can be obtained. It can, however, be reached, alternatively, by letting candidates 3 and 4 exchange their jobs.

A problem which is similar to that just discussed must be solved if, instead of rating numbers, we are merely told whether a candidate is, or is not, fit for a job.

This problem can be described by using only merit numbers 1 (if suitable) and 0 (if not suitable). Also, in this case we can ask a related question: is it possible to fill all vacancies with suitable candidates?

The latter problem can be reformulated, and connected with an interesting combinatorial problem, as follows.

Let the candidates be denoted by $i = 1, 2, \ldots, n$, and the jobs by $j = 1, 2, \ldots, m$, where $m \leqslant n$ (otherwise the answer is certainly in the negative). We call S_j the set of the indices of all those candidates who are suitable for job j, and we want to choose one suitable candidate for each job, different from all other candidates chosen for another job.

If we call an individual in a set one of its "representatives," then we wish to find a "system of distinct representatives" and our problem is reduced to that of the existence of such a system. Philip Hall has obtained a theorem about this question,* and we are now going to derive it by linear programming methods, using properties of duality.

We allocate candidates to each of the m jobs in such a way that the merit numbers (1 or 0) of those allocated within their jobs add up to as large a number as possible. This is a simple Transportation Problem. If the maximum is m, then every job has been given to a suitable candidate. If this is impossible, then the maximum will be smaller than m. We are thus looking for a criterion whether that maximum is m or less than m.

The problem can be written:

$$\text{Maximize} \sum_i \sum_j a_{ij} x_{ij}$$

subject to $\sum_j x_{ij} \leqslant 1$ for all i, and $\sum_i x_{ij} \leqslant 1$ for all j, $x_{ij} \geqslant 0$, where a_{ij} is 1 or 0, dependent on whether candidate i is suitable or not suitable for job j. The first inequality comes from the fact that not every candidate is necessarily allocated to some job, and the second from the fact that if the maximum is less than m anyway, some jobs might just as well remain unfilled.

The dual of this problem is

$$\text{Minimize} \sum_{i=1}^{n} u_i + \sum_{j=1}^{m} v_j$$

for non-negative u_i and v_j, subject to $u_i + v_j \geqslant a_{ij}$ for all pairs (i, j).

If the u_i and v_j were independent of one another, then we would make all of them equal to zero. However, this would contradict

* See [24].

$u_i + v_j \geqslant a_{ij}$ whenever $a_{ij} = 1$. In such a case either u_i or v_j will have to be 1.

Imagine that the dual problem has been solved, and that the solution can be described as follows:

	Number of u_i	Number of v_j
variables equal to		
1	t	$m - k$
0	$n - t$	k
Total	n	m

Then $\sum_i u_i + \sum_j v_j = t + m - k$. If $t = k$, then the minimum of this objective function, and hence the maximum of the objective function of the original problem, is m: all vacancies can be filled by suitable candidates. But if the minimum is smaller than m (it can not be larger because it equals the maximum of the primal problem), then some vacancies will remain unfilled, and we want to find a criterion when this is so.

$$\sum_i u_i + \sum_j v_j < m \text{ means that } k > t$$

But k is the number of those v_j which are 0 in the optimal solution of the dual, so that every time when a_{ij} was 1 for one of these j, u_i was made 1 instead of the v_j. Since only t of the u_i were made unity, there could not have been more than t different i connected with these j, when a_{ij} was unity. In other words, there must be k jobs (those for which $v_j = 0$) for which less than k individuals were altogether suitable.

This is the case when it is impossible to fill all the vacancies. It follows that if, for each k, any k of the sets S_j contain at least k items between them, then a system of distinct representatives can be constructed. This is a sufficient condition, and it is obvious that this condition is also necessary. We have thus proved Philip Hall's Theorem.*

To illustrate our result by examples:
If the rating numbers are as follows:

	Jobs			
Candidates	1	1	0	0
	0	0	1	1
	0	1	0	1
	1	0	1	0

* The essence of this derivation is given in [28].

then we can arrange the following allocation:

(1, 1), (2, 4), (3, 2), (4, 3)

(this scheme is not the only possible one, by the way) and we find, on inspection, that any 2 (or 3, or all 4) columns contain at least 2 (or 3 or 4) unit values between them. On the other hand, if the rating numbers were

	Jobs			
	1	1	1	1
Candidates	0	0	1	1
	0	0	0	1
	0	0	1	0

then clearly such an allocation is impossible: only one, and the same, candidate is suitable for job 1 and 2.

19 Selling Periodicals

A publisher and a newsagent come to the following arrangement. The newsagent takes from the publisher a certain number of copies of a monthly journal, paying 6p. per copy. He sells them at 15p. per copy. After two weeks he can return unsold copies, to the publisher, receiving 4p. for each, and he can also buy new copies, again at 6p. At the end of the month he can return all unsold copies, this time for 2p. each.

Let it be assumed that the demand for copies during the first two weeks will be

10	with probability	$\frac{1}{4}$
14	,, ,,	$\frac{1}{2}$
18	,, ,,	$\frac{1}{4}$

During the second half of the month it will be

4	with probability	$\frac{3}{4}$
5	,, ,,	$\frac{1}{4}$

How many copies should the newsagent buy from the publisher to begin with? What should he do after two weeks dependent on how many copies are left?

The newsagent will not require less than 14, or more than 23 altogether. We consider first the possible situations after two weeks, and we shall then work backwards from our conclusion to the beginning of the month.

If no copies are left after two weeks, then we order either 4 or 5 copies to satisfy the expected demand. If we order 4, then we pay 24 and sell them for 60, thus making a profit (during the second half of the month) of 36. If we order 5, then the profit is either 45 (if all 5 are sold) or 32 (if 4 are sold, since then the profit on these is 36, but the loss on the unsold copy is 6 less 2, i.e. 4).

More generally, assume that $b \leqslant 4$ copies remain unsold. We then order either $4 - b$, or $5 - b$ further copies. If the actual demand is 4, then we gain, respectively,

$$60 - 6(4 - b) \quad \text{or} \quad 60 - 6(5 - b) + 2$$

while, if the demand is 5, we gain

$$60 - 6(4 - b) \text{ as before, or } \quad 75 - 6(5 - b)$$

Numerically, we obtain the following table, where the first entry in each cell refers to the gain when the demand is 4, and the entry after the comma to that when the demand is 5 copies:

Unsold	Order					
copies	0	1	2	3	4	5
0					36,36	32,45
1				42,42	38,51	
2			48,48	44,57		
3		54,54	50,63			
4	60,60	56,69				

If there are 5 or more copies left, then we can return the excess over 4, or that over 5. There will not be more than 13 copies left because we do not buy more than 23 to begin with, and at least 10 will have been sold, according to our assumptions. The extension of the table above, to deal with returns, is as follows:

Unsold	Return									
copies	9	8	7	6	5	4	3	2	1	0
5									64,64	62,75
6								68,68	66,79	
7							72,72	70,83		
8						76,76	74,87			
9					80,80	78,91				
10				84,84	82,95					
11			88,88	86,99						
12		92,92	90,103							
13	96,96	94,107								

Dependent on how many copies have been sold during the first two weeks, we make then that decision which makes the expected further gain largest. This implies the following consideration.

The demands 4 and 5 have, respectively, probabilities $\frac{3}{4}$ and $\frac{1}{4}$, so that, if no copy has remained unsold, it is preferable to order 4 rather than 5 because the expected gain from 4 copies, i.e.

$$\tfrac{3}{4} \text{ times } 36 + \tfrac{1}{4} \text{ times } 36 = 36$$

is larger than

$$\tfrac{3}{4} \text{ times } 32 + \tfrac{1}{4} \text{ times } 45 = 35\tfrac{1}{4}$$

Generally, we find that as long as not more than 4 copies are left, the best procedure is to buy copies so as to have 4 altogether, while if 5 or more are left it is best to return the excess over 5 (not over 4). To illustrate this latter case, assume that 8 copies are left after two weeks. If we returned 4, the expected gain would be 76. If we returned only 3, the expected gain would be

$$\tfrac{3}{4} \text{ times } 74 + \tfrac{1}{4} \text{ times } 87 = 77\tfrac{1}{4}$$

which is higher.

Having thus found the optimal decision to be made after two weeks, dependent on the number of copies still unsold, we turn now to the initial decision, concerned with buying 14, or 15, or 16, etc., . . ., or 23 copies.

The probabilities of demand during the first two weeks produce the following possibilities:

Initial purchase	*Cost*	*After two weeks there remain*				
					having sold	at
14	84	4	0	0	14	210
15	90	5	1	0	15	225
16	96	6	2	0	16	240
17	102	7	3	0	17	255
18	108	8	4	0	18	270
19	114	9	5	1	18	270
20	120	10	6	2	18	270
21	126	11	7	3	18	270
22	132	12	8	4	18	270
23	138	13	9	5	18	270
		with probability $\frac{1}{4}$ having sold 10 at 150 altogether	with probability $\frac{1}{2}$ having sold 14 at 210 altogether	with probability $\frac{1}{4}$		

Hence, if the newsagent acquires, for instance, 14 copies, then he will sell 10 with probability $\frac{1}{4}$ and will then have a residue of 4 copies. By virtue of the optimal decision to be made on this basis, he will not order or return any, and he will have a further gain of 60. Adding to this his gain during the first two weeks, namely 10 times 15 less 84, his total gain will be 126. Similarly, looking at the second and third column above, he will have a gain of 14 times 15 less 84, i.e. 126, during the first half of the month. Together with his gain of 36 during the second half, he will have 162, with probability $\frac{1}{2} + \frac{1}{4} = \frac{3}{4}$. Thus his expected gain, on buying 14 copies to begin with, is

$$\tfrac{1}{4} \text{ times } 126 + \tfrac{3}{4} \text{ times } 162 = 153$$

In the same way we obtain the figures of the following table:

Initial purchase	*Total gain with probability* $\frac{1}{4}$	$\frac{1}{2}$	$\frac{1}{4}$	*Expected total*
14	126	162	162	153
15	$125\frac{1}{4}$	162	171	155·0625
16	$123\frac{1}{4}$	162	180	156·8125
17	$121\frac{1}{4}$	162	189	158·5625
18	$119\frac{1}{4}$	162	198	160·3125
19	$117\frac{1}{4}$	$161\frac{1}{4}$	198	159·4375
20	$115\frac{1}{4}$	$159\frac{1}{4}$	198	157·9375
21	$113\frac{1}{4}$	$157\frac{1}{4}$	198	156·4375
22	$111\frac{1}{4}$	$155\frac{1}{4}$	198	154·9375
23	$109\frac{1}{4}$	$153\frac{1}{4}$	$197\frac{1}{4}$	153·25

The total expected gain is largest for an initial purchase of 18 copies.

This dynamic programming approach (cf. [5]) is used in [29].

20 The Taxi-driver's Dilemma

A taxi-driver can work in either of two places, A and B, and in either he can employ two methods when free: he can cruise (I), or he can go to a taxi-stand (II).

If he is in A and cruises, then there is a probability of 1/2 that he will take his next passenger to somewhere in A, for a reward of 10, and a probability of 1/2 that he takes him to B, for a reward of 4. If he goes to a stand, then the respective probabilities are 3/4 (to A) and 1/4 (to B), for a return of 8, or 2, respectively.

On the other hand, if he is in B, then he has a probability of 1/4 of taking his next passenger to somewhere in A for a return of 10 or to somewhere in B for a return of 2, assuming that he cruises, while if he goes to a stand, his probabilities and returns are 1/8 for 8 (to A) and 7/8 for 4 (to B).

Having reached his destination, he looks for another fare, and we want to determine the best succession of choices, if there are (*a*) two more or (*b*) a very great number of more fares to come.

We exhibit the relevant information as follows:

In A

	Probabilities		Rewards	
Doing	to A	to B	to A	to B
I	0·5	0·5	10	4
II	0·75	0·25	8	2

hence the expected return is

with I	7
with II	6·5

In B

	Probabilities		Rewards	
Doing	to A	to B	to A	to B
I	0·25	0·75	10	2
II	0·125	0·875	8	4

hence the expected return is

with I	4
with II	4·5

It follows that if he is in A, he should cruise, and if he is in B, he should go to a stand, provided he is only thinking of a return from the next fare.

Now imagine that the taxi-driver is thinking of the next two fares, and wants to maximize his expected income from these. Again, his decision depends on where he is.

If he is in A and cruises, he can expect a return of 10 with probability 0·5; and then to find himself again in A, where he will cruise and have a further expected income of 7, or a return of 4 with probability 0·5; and then to find himself in B, where he will go to a stand and have a further expected income of 4·5. He must therefore do the following computations:

In A

	Expected reward
Cruising:	$0{\cdot}5(10 + 7) + 0{\cdot}5(4 + 4{\cdot}5) = 12{\cdot}75$

Similarly, he must work out what to expect if he goes to a stand:

Going to a stand: $0{\cdot}75(8 + 7) + 0{\cdot}25(2 + 4{\cdot}5) = 12{\cdot}875$

The second result is larger; therefore, if he is in A, he should go to a stand, get his fare, and then do the best for the one remaining fare, dependent on where he finds himself.

For B, he works out the following:

In B

	Expected reward
Cruising:	$0{\cdot}25(10 + 7) + 0{\cdot}75(2 + 4{\cdot}5) = 9{\cdot}125$
Going to a stand:	$0{\cdot}125(8 + 7) + 0{\cdot}875(4 + 4{\cdot}5) = 9{\cdot}3125$

This time, once more, the second result is larger; therefore he should also go now to a stand, if he is in B, provided he thinks that he has two fares outstanding.

If he has three fares outstanding, then he will again work out what to do next, under the assumption that, wherever he is next time, he will make the best choice which we have just worked out for the case of two more fares. The computations for this case are:

In A

	Expected total reward
Cruising:	$0{\cdot}5(10 + 12{\cdot}875) + 0{\cdot}5(4 + 9{\cdot}3125) = 18{\cdot}1$
Going to a stand:	$0{\cdot}75(8 + 12{\cdot}875) + 0{\cdot}25(2 + 9{\cdot}3125) = 18{\cdot}5$

In B

	Expected total reward
Cruising:	$0{\cdot}25(10 + 12{\cdot}875) + 0{\cdot}75(2 + 9{\cdot}3125) = 14{\cdot}2$
Going to a stand:	$0{\cdot}125(8 + 12{\cdot}875) + 0{\cdot}875(4 + 9{\cdot}3125) = 14{\cdot}3$

Thus he should, wherever he is, go to a stand if he expects to have three more fares. If he expects more, then he would make similar computations, by this "Value Iteration Method" (cf. [26]).

W.S. Jewell has studied the case when the times between transitions are random, and the reward depends on these intervals as well—as would be the case if one takes, realistically, the length of any taxi-drive also into account. He speaks in this case of a Markov Renewal Process.*

P. Wolfe and G. B. Dantzig have pointed out, in [39], that the problem of many more fares to come, which is also considered in Howard's book [26], can also be tackled by linear programming, and Derman and Klein (see [13]) have shown that this is also true if that number is small. (See also the bibliography in the last-mentioned paper.)

* In [40].

21 Betting on Race-courses

Bookmakers offer you the following odds for the next two races:

Simplex Stakes		Dynamic Derby	
Fair Lady	3:2	Red Heat	6:1
Sound of Music	8:3	Black Jack	7:2
Oklahoma	5:2	Yellow Peril	2:1
Desert Song	7:1	White Wash	5:2

You are willing to bet, provided you are certain to win. Have you got a chance?

Let the odds be $a_i:1$, which means that if you stake x_i on horse i, and it wins, then you win a_ix_i, while otherwise you lose your stake x_i. Hence if you stake a proportion x_i of your total bet on horse i, then

$$x_1 + x_2 + x_3 + x_4 = 1$$

and you get, if horse 1 wins,

$$a_1x_1 - x_2 - x_3 - x_4 = (a_1 + 1)x_1 - 1$$

and, generally, if horse i wins, you get

$$(a_i + 1)x_i - 1$$

We want the smallest of these gains to be as large as possible. Let us call this smallest value v, say, then

$$(a_i + 1)x_i - 1 \geqslant v \qquad (\text{for } i = 1, 2, 3, 4)$$

We introduce slack variables y_i, and eliminate v (which is not sign-restricted), so that we have the system

$$\begin{aligned}&x_1 + x_2 + x_3 + x_4 = 1\\&(a_1 + 1)x_1 - y_1 - 1 = (a_2 + 1)x_2 - y_2 - 1\\&= (a_3 + 1)x_3 - y_3 - 1 = (a_4 + 1)x_4 - y_4 - 1\end{aligned}$$

We want to maximize v, or equivalently $4v$, which is the sum of the last four expressions

$$4v = \sum_{i=1}^{4} (a_i + 1)x_i - \sum_{i=1}^{4} y_i - 4 = B, \text{ say}$$

If we write $x_i = k/(a_i + 1)$, then the constraints are satisfied with $= 0$, provided

$$k = 1 \bigg/ \sum_{i=1}^{4} \frac{1}{a_i + 1}$$

Expressing the x_i in terms of the non-basic y_i, we have

$$x_i = \frac{k}{c_i} + \frac{y_i}{c_i} - \frac{k}{c_i} \sum_{i=1}^{4} \frac{y_i}{c_i}$$

where we have written c_i for $a_i + 1$, and the value of B is

$$4\left[k - 1 - k \sum_{i=1}^{4} \frac{y_i}{c_i}\right]$$

The coefficients of all y_i are negative, and hence we have obtained the maximum of v, which is $(k - 1)$. The best distribution of the total stake is given by the x_i above. If we bet in this way, then whichever horse wins, the outcome for us is the same.

However, does this mean a gain? In algebraic terms:

$$\text{is } k > 1, \text{ i.e. is } \sum_{i=1}^{4} \frac{1}{a_i + 1} < 1?$$

For the Simplex Stakes we have

$$\frac{1}{2{\cdot}5} + \frac{1}{11/3} + \frac{1}{3{\cdot}5} + \frac{1}{8} = 0{\cdot}4 + 0{\cdot}2727 + 0{\cdot}2857 + 0{\cdot}125 = 1{\cdot}0834$$

In this case it is not worth betting at all, since the best we can get is to win $1/1{\cdot}0834 - 1$, which is a loss.

In the Dynamic Derby we compute

$$\frac{1}{7} + \frac{1}{4{\cdot}5} + \frac{1}{3} + \frac{1}{3{\cdot}5} = 0{\cdot}1429 + 0{\cdot}2222 + 0{\cdot}3333 + 0{\cdot}2857$$
$$= 0{\cdot}9841$$

so that in this race we have a certain win. We have to bet, on the four horses, in the proportion

$$1/7 : 1/4{\cdot}5 : 1/3 : 1/3{\cdot}5 \quad \text{or} \quad 9/62 : 7/31 : 21/62 : 9/31.$$

Then, whichever horse wins, we get $1/0{\cdot}9841 - 1 = 1/62$ times the total stake as a profit. (This is not much, but sure.) For instance, if Red Heat wins, then we win $6 \times 9/62$ and lose

$$\frac{7}{31} + \frac{21}{62} + \frac{9}{31} = \frac{53}{62}$$

if White Wash wins, then we win $\frac{5}{2} \times \frac{9}{31}$ and lose

$$\frac{9}{62} \times \frac{7}{31} + \frac{21}{62} = \frac{22}{31}$$

In each case, and similarly for the other horses, the difference is 1/62.

This way of judging the result of judicious betting is mentioned in [1], p. 531.

22 Paradoxes

Consider the following problem. A farmer can grow three types of produce, say A, B, and C. For this he needs, per unit of produce, respectively

for	A	B	C	
Units of acreage	1	2	1	
Units of labour	3	1	2	
and earns a profit of	1	3	2	units of money

Given that he has 4 units of acreage and 12 units of labour at his disposal, he wishes to maximize his total profit.

What happens if he uses all his acreage and 9 units of labour?

Formulating this as a linear programming problem, and introducing x_1, x_2, and x_3 as the numbers of units of products A, B, and C, the farmer wants to maximize

$$B = x_1 + 3x_2 + 2x_3$$

subject to

$$\begin{aligned} x_1 + 2x_2 + x_3 &= 4 \\ 3x_1 + x_2 + 2x_3 &= 9 \qquad x_i \geqslant 0 \end{aligned}$$

The solution to this problem is

$$x_1 = 1, \quad x_2 = 0, \quad x_3 = 3, \text{ and } B = 7$$

If he uses all his 12 units of labour, and maximizes B as above, subject to

$$\begin{aligned} x_1 + 2x_2 + x_3 &= 4 \\ 3x_1 + x_2 + 2x_3 &= 12 \qquad x_i \geqslant 0 \end{aligned}$$

then the solution is

$$x_1 = 4, \quad x_2 = 0, \quad x_3 = 0, \text{ and } B = 4,$$

thus he gets *less* for using *more* labour. This is our first paradox.

If he used only 5 units of labour, he would obtain, from $x_1 = 0$,

$x_2 = 1, x_3 = 2$, a profit of 7, as much as from 9 units, but again more than from 12 units of labour.

We want to see more clearly what happens.

Generally, let us use L units of labour, as the right-hand side of the second constraint. The basic solutions of the constraints are then:

(1) $x_1 = 0, \quad x_2 = (8 - L)/3, \quad x_3 = (2L - 4)/3,$
$B = (16 + L)/3$ feasible when $2 \leqslant L \leqslant 8$, so that $6 \leqslant B \leqslant 8$.

(2) $x_1 = L - 8, \quad x_2 = 0, \quad x_3 = 12 - L, \quad B = 16 - L$
feasible when $8 \leqslant L \leqslant 12$, so that $4 \leqslant B \leqslant 8$.

(3) $x_1 = (2L - 4)/5, \quad x_2 = (12 - L)/5, \quad x_3 = 0,$
$B = (32 - L)/5$ feasible when $2 \leqslant L \leqslant 12$, so that $4 \leqslant B \leqslant 6$, but in this case B is a minimum!

The following diagram illustrates the situation:

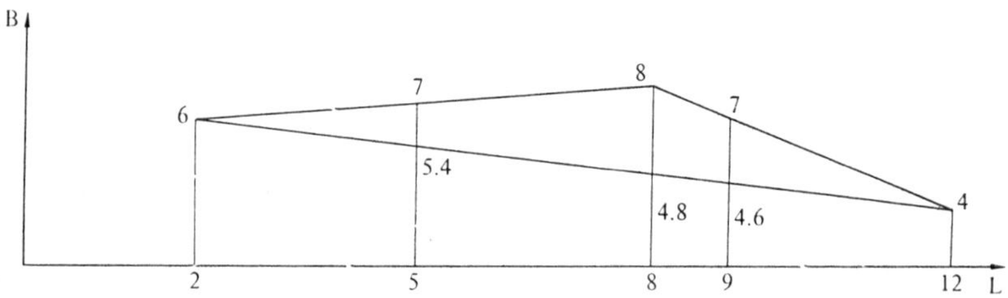

We now see how the paradox arose. We see also that while for $L = 9$ the largest B we can obtain is 7, it is not true that to obtain 7, we must use at least $L = 9$. $L = 5$ would do equally well. (In fact, $L = 9$ is the *largest* value to obtain 7.)

On the other hand, while for $L = 5$ the most we can obtain is 7, it is true that to obtain 7, we must use at least $L = 5$.

Because of this later property, we call the set of points

$$2 \leqslant L \leqslant 8 \qquad B = (16 + L)/3$$

an *efficient* set.

It we wanted to use $L < 2$, or $L > 12$, then the two equality constraints could not be solved in non-negative values.

Clearly, the paradox arises because we insist on using all of the available facilities. In practise we should not use any value of L if we obtain the same benefit from using less.

Therefore we consider the system of *inequalities:*

$$x_1 + 2x_2 + x_3 \leqslant 4$$
$$3x_1 + x_2 + 2x_3 \leqslant L$$

i.e. with slack variables.

$$x_1 + 2x_2 + x_3 + x_4 = 4$$
$$3x_1 + x_2 + 2x_3 + x_5 = L$$

In this case, for $0 \leqslant L \leqslant 2$, B is maximized by $x_2 = L$; for $2 \leqslant L \leqslant 8$ the answer is, of course, the same as for the efficient set in the case of equality constraints; and for $8 \leqslant L \leqslant 12$ the maximum of B remains at 8, with $x_3 = 4$.

The dual to the problem, with inequalities, is

$$\text{minimize } 4y_4 + Ly_5 = C \text{ say}$$
$$\text{subject to} \quad y_4 + 3y_5 \geqslant 1$$
$$2y_4 + y_5 \geqslant 3$$
$$y_4 + 2y_5 \geqslant 2$$
$$y_4, y_5 \geqslant 0$$

If we take, for instance, $L = 12$, then the answer is

$$y_4 = 2, \quad y_5 = 0, \quad C = 8$$

On the other hand, the dual to the system with equality constraints, is the same, except that the y_j are not sign-restricted. The answer is then, for $L = 12$,

$$y_4 = 4, \quad y_5 = -1, \quad C = 4$$

We notice that one of the variables is negative. It follows that, if they were sign-restricted, C (which we want to minimize) would be larger, and so would B, the objective function of the primal (which we want to maximize). In other words, if the primal is written in *inequalities* (and $L = 12$), then a better result can be obtained.

This is true for all values of L, which produce a negative variable in the dual, such as $y_5 = -1$ above. Now the optimal solutions of the dual of the system in equations, with general L, between 8 and 12, is

$$y_4 = \tfrac{1}{2}L - 2 \qquad y_5 = 2 - L/4$$

so that, as long as L exceeds 8, y_5 will be negative, and it is better not to use that much of L as is available.

We turn now to another problem, one of Transportation.
Let there be available, at A and at B, $a_1 = 6$ and $a_2 = 1$,
and let it be required to deliver at C, $b_1 = 4$, and at D, $b_2 = 3$.
The cost table is assumed to be

	C(4)	D(3)
A(6)	10	3
B(1)	2	5

The smallest total cost, obtained when

$$\text{AC} = 3, \quad \text{AD} = 3, \quad \text{BC} = 1$$

is 41.

Suppose now that the availability at A is increased to $a_2 = 2$, and the requirement at D is increased to $b_2 = 4$. The best plan is

$$\text{AC} = 2, \quad \text{AD} = 4, \quad \text{BC} = 2$$

and the total cost is 36.

Thus, although we send more through the same system (8 against 7), the total cost is now reduced (from 41 to 36). This is the so-called *Doig Paradox* (after Alison Doig, who discovered it). It arises because, if more is available at B, we can make better use of the cheap route BC.

We could have recognized this possibility by looking again at the dual. The optimal dual variables (shadow costs) are then

		0
		-8
10	3	

The constraints of the Transportation Problem are equations, and hence the dual variables u_i and v_j of the dual problem

$$\begin{aligned} &\text{maximize } a_1u_1 + a_2u_2 + b_1v_1 + b_2v_2 \\ &\text{subject to} \quad u_1 + v_1 \leqslant 10 \qquad u_1 + v_2 \leqslant 3 \\ &\qquad\qquad\quad u_2 + v_1 \leqslant 2 \qquad u_2 + v_2 \leqslant 5 \end{aligned}$$

are not sign-restricted.

If we have a pair $u_i + v_j < 0$, and if we increase a_i and b_j by the same amount $c(> 0)$, then the objective function will be changed by $c(u_i + v_j)$, a negative amount, and the minimum cost of transportation will be reduced. In our example we have $u_2 + v_2 = -5$, so that the addition of 1 to a_2 and to b_2 reduced the total cost by 5. This can go on as long as the basic variables (those referring to routes which we actually use) remain unaltered. In our example we can continue until, in the optimal solution:

	4	$2 + a_2$
6	$4 - a_2$	$2 + a_2$
a_2	a_2	

we have $a_2 = 4$. The cost is then 26. If we added one more unit to a_2 and b_2, then we would have the optimal solution

	4	7
6		6
5	4	1

which would cost 31, again more than 26.

Appendix

The Simplex Method

The Simplex Method is an iterative procedure for solving linear programming problems, which gives the final answer after a finite number of steps.

Let it be required to maximize (to minimize) the "objective function"

$$x_0 = \sum_{i=1}^{n} b_i x_i$$

subject to the "constraints" $\sum_{i=1}^{n} a_{ij} x_i = c_j \qquad (j = 1, 2, \ldots, m)$

and $\qquad x_i \geqslant 0$

(Conventionally, we write B for x_0 if we want to stress the fact that we maximize, and C when we minimize.)

If some constraints are in the form

$$\sum a_i x_i \leqslant c \qquad \text{or} \qquad \sum a_i x_i \geqslant c$$

then we introduce a non-negative "slack" variable y and write

$$\sum a_i x_i + y = c \qquad \text{or} \qquad \sum a_i x_i - y = c$$

It might also be convenient to introduce "artificial" variables z, thus

$$\sum a_i x_i + z = c \qquad \text{or} \qquad \sum a_i x_i - z = c$$

and to add Mz to the objective function when we minimize, or $-Mz$ when we maximize, where M is a large number. Such an artificial variable will be zero in the final answer, unless the constraints are contradictory.

At any stage during the iterations we express m "basic" variables in terms of $(n - m)$ "non-basic" variables, thus:

$$x_s + \sum_t z_{st} x_t = z_{s0} \geqslant 0$$

for all subscripts s of the basic variables, where the summation extends over all subscripts t of the non-basic variables. The objective function, also in terms of the non-basic variables, has the form

$$x_0 + \sum_t z_{0t} x_t = z_{00}$$

We write this in tableau form as follows:

		b_{t_1}		$b_{t_{n-m}}$	
		x_{t_1}	. . . .	$x_{t_{n-m}}$	
b_{s_1}	x_{s_1}	$z_{s_1t_1}$		$z_{s_1t_{n-m}}$	z_{s_10}
b_{s_m}	x_{s_m}	$z_{s_mt_1}$		$z_{s_mt_{n-m}}$	z_{s_m0}
	x_0	z_{0t_1}		$z_{0t_{n-m}}$	z_{00}

The b values are sometimes added to make it easier to establish that

$$bz_{s_1t} + bz_{s_2t} + \ldots + bz_{s_mt} - b_t = z_{0t}$$

for all subscripts t of the non-basic variables (and, of course, $b_{s_1}z_{s_10} + \ldots + b_{s_m}z_{s_m0} = z_{00}$ must also hold).

If we want to maximize (minimize), then the final tableau has been reached when all $z_{0t} \geqslant 0$ ($z_{0t} \leqslant 0$). Otherwise we choose some z_{0t} which has the wrong sign and find that s for which

$$z_{s0}/z_{st}$$

is smallest, while $z_{st} > 0$. The z_{st} for which these conditions hold is called the "pivot," and is indicated by an asterisk below. The tableau is transformed into the next as follows:

From

	x_t	x_v	
x_s	z_{st}^*	z_{sv}	z_{s0}
x_u	z_{ut}	z_{uv}	z_{u0}
x_0	z_{0t}	z_{0v}	z_{00}

into

	x_s	x_v	
x_t	$1/z_{st}$	z_{sv}/z_{st}	z_{s0}/z_{st}
x_u	$-z_{ut}/z_{st}$	$z_{uv} - z_{sv}z_{ut}/z_{st}$	$z_{u0} - z_{s0}z_{ut}/z_{st}$
	$-z_{0t}/z_{st}$	$z_{0v} - z_{sv}z_{0t}/z_{st}$	$z_{00} - z_{s0}z_{0t}/z_{st}$

where x_v is a non-basic variable which remains non-basic, x_u a basic variable which remains basic, and x_s and x_t are exchanging their roles.

We may also start with—or meet at a later stage—a tableau where all $z_{0t} \geqslant 0$ (when we wish to maximize) or all $z_{0t} \leqslant 0$ when we want to minimize (in other words, where they have already their desired sign), but not all $z_{s0} \geqslant 0$, some being "unfeasible." We can then proceed by the Dual Simplex Method and choose a z_{s0} which is negative. The pivot is that z_{st} which is negative and for which the absolute value $|z_{0t}/z_{st}|$ is smallest. The rules of tableau transformation are the same as before.

Duality

Consider the two systems which are "dual" to one another:

Maximize	Minimize
$\sum_{i=1}^{n} b_i x_i$	$\sum_{j=1}^{m} c_j y_{n+j}$
subject to	subject to
$\sum_{i=1}^{n} a_{ij} x_i + x_{n+j} = c_j$	$\sum_{j=1}^{m} a_{ij} y_{n+j} - y_i = b_i$
$(j = 1, \ldots, m)$	$(i = 1, \ldots, n)$
and	and
x_i, x_{n+j} non-negative	y_{n+j}, y_i non-negative
$(i = 1, \ldots, n;\ j = 1, \ldots, m)$	$(j = 1, \ldots, m;\ i = 1, \ldots, n)$

where we assume that the b_i and the c_j are positive. We can then start solving the maximizing problem by the Simplex Method, and the minimizing problem by the Dual Simplex Method.

The initial tableaux are

	x_1	. . . . .	x_n	
x_{n+1}	a_{11}	. . . . .	a_{n1}	c_1
		. . . . .		
x_{n+m}	a_{1m}	. . . . .	a_{nm}	c_m
	$-b_1$	. . . . .	$-b_n$	0

	y_{n+1}	. . . . .	y_{n+m}	
y_1	$-a_{11}$	. . . . .	$-a_{1m}$	$-b_1$
		. . . . .		
y_n	$-a_{n1}$	. . . . .	$-a_{nm}$	$-b_m$
	$-c_1$	. . . . .	$-c_m$	0

As we proceed, we can at each stage choose the pivots in such a position that the relationship of the two tableaux is the same as above: the entries within the tableau are transposed, and the signs changed, except for the values in the positions of the b_i above, which retain their signs. We shall then reach the final tableaux at the same stage. In particular, the two optimal values of the objective functions will be equal. Consequently, we may at every stage turn from one tableau to its dual and continue with the alternative method, whenever this is convenient. We solve in any case both problems simultaneously, and from the final solution of one of them, that of its dual can be derived.

Inverse Matrix Method

This method, also called the Revised Simplex Method, is a version of the original Simplex Method which has proved useful in automatic computation. Let our statement of the problem, in matrix form, be

x_1	.	.	.	.	x_n	x_0	
a_{11}	.	.	.	.	a_{n1}	0	c_1
—	—	—	—	—	—	—	
a_{1m}	.	.	.	.	a_{nm}	0	c_m
$-b_1$	.	.	.	.	$-b_n$	1	0

At each stage, we use a square matrix, the inverse of that submatrix of the original one above, which consists of the columns of the current basic variables (including x_0, which remains basic throughout). We also add, as a last column, the current values of the basic variables, thus:

x_{s1}	$r_{s_1 1}$	.	.	.	.	$r_{s_1 m}$	0	$z_{s_1 0}$
—	— — — —							
x_{s_m}	$r_{s_m 1}$	.	.	.	.	$r_{s_m m}$	0	$z_{s_m 0}$
x_0	r_{01}	.	.	.	.	r_{0m}	1	z_{00}

In order to establish whether we have reached the final answer, we pre-multiply the columns of the original matrix by the last row of the inverse matrix, i.e. we compute

$$r_{01}a_{j1} + \ldots + r_{0m}a_{jm} - b_j$$

for all $j = 1, \ldots, n$. If j is the subscript of any of the basic variables $x_{s_1}, \ldots, x_{s_m}$, then the result must be zero. The other results are the values z_{0j} of the current tableau of the Simplex Method. If we find a value with the "wrong" sign, say in the column of the non-basic variable x_t, then we know that we must continue.

We construct the complete column of x_t in the current Simplex Tableau as follows:

$$r_{s_1 1}a_{t1} + \ldots + r_{s_1 m}a_{tm} \qquad = z_{s_1 t}$$
$$- - -$$
$$r_{s_m 1}a_{t1} + \ldots + r_{s_m}a_{tm} \qquad = z_{s_m t}$$
$$r_{01}a_{t1} + \ldots + r_{0m}a_{tm} - b_t = z_{0t}$$

(This last value has already been worked out.) The pivot $z_{s_h t}$ is then determined in the usual way, and the next inverse matrix and its column of constants found by pre-multiplying the last tableau by the modified identity matrix (also called elementary matrix) which is the identity matrix (all diagonal values 1, and all other values 0), but having its *h*-th column replaced by

$$-z_{s_1 t}/z_{s_h t}$$
$$-z_{s_2 t}/z_{s_h t}$$
$$\cdot$$
$$\cdot$$
$$1/z_{s_h t}$$
$$-z_{s_{h+1} t}/z_{s_h t}$$
$$\cdot$$
$$\cdot$$
$$-z_{0t}/z_{s_h t}$$

(The denominator $z_{s_h t}$ is, of course, the pivot.)

N.B. The result of pre-multiplying an *r*-by-*s* matrix (b_{jk}) by a *t*-by-*r* matrix (a_{ij}) is a *t*-by-*s* matrix $(\sum a_{ij}b_{jk})$, summation over *j*.

Decomposition Algorithm

This algorithm is a modification of the Simplex Method, or the Inverse Matrix Method, which is particularly useful if a problem can be formulated as follows:

Maximize $x_0 = \sum_{i=1}^{n} b_i x_i$, with non-negative variables, subject to

$$\sum_{i=1}^{n} a_{ij}x_i = c_j \qquad (j = 1, \ldots, m)$$

$$\sum_{i=1}^{n_1} a_{ik}x_i = c_k \qquad (k = 1, \ldots, m_1)$$

$$\sum_{i=n_1+1}^{n_2} a_{il}x_i = c_l \qquad (l = 1, \ldots, m_2)$$

$$\sum_{i=n_{t-1}+1}^{n_t} a_{ir}x_i = c_r \qquad (r = 1, \ldots, m_t)$$

We refer to the first m equations, which contain all n variables, as "linking" equations, and to the other sets of constraints, which contain several non-overlapping subsets of the variables, as "block" equations. We shall also be interested in the basic feasible solutions (vertices of the feasible regions) of the block equations.

We consider, instead of the original system of constraints, the following:

$$\sum_{i=1}^{n} a_{ij}x_i = c_j \qquad (j = 1, \ldots, m) \text{ (the linking equations)}$$

$$\sum \lambda_i^1 = 1$$

$$\sum \lambda_i^t = 1$$

where the $\lambda_i^1, \ldots, \lambda_i^t$ are non-negative variables, as many in number as there are basic feasible solutions in the 1st, . . ., t-th block equations. At every stage of the Inverse Matrix Method applied to this system, there will be some x and/or some λ variables in the basis. To decide whether a further iterative step is called for, we solve the following ancillary linear programming problem:

$$\text{minimize } r_{01}(A_1x) + r_{02}(A_2x) + \ldots + r_{0m}(A_mx) - (Bx)$$

where A_ix is the sum of terms of variables appearing in a selected block in the i-th constraint of that block, and (Bx) is the sum of terms of those variables in the objective function.

We minimize, subject to the constraints of the selected block. If the minimum is non-negative, then we turn to another block, if any. If we do not find any negative value after having treated all the blocks, then we know that we have found the final answer. If we find a negative value, then we introduce a λ, which corresponds to the vertex where we have found the minimum. If the basic feasible solution of the ancillary problem is given by $x^1, \ldots, x^g$ (values of the variables in the block considered), then we imagine that we have, in the modified system of constraints, the following coefficients of the λ to be introduced:

$$\sum_{i=1}^{g} a_{ij}x^i \qquad (j = 1, \ldots, m)$$

1 in the constraint corresponding to the block of that λ
0 in the constraints of all the other blocks.

We then transform this column according to the procedure of the Inverse Matrix Method, find a pivot, and exchange λ for one of the current basic variables.

Having found the final answer, we must translate the basic λ variables into x variables. We consider the former as weights of the vertices in the feasible region of their respective blocks, and determine the x variables, in that feasible region, of their centre of gravity.

Integer Programming

When it is required that all variables (pure integer programming) or certain specified variables (mixed integer programming) must have integer values, then we can, for instance, apply methods due to R. E. Gomory (cf. [22] and [23]).

To begin with, we ignore the condition of integrality and apply the Simplex Method. If, in the final tableau, the variables which ought to be integers are not all such, then we add a further row of constraints to the tableau.

Assume that a row in the final Simplex Tableau is

	x_{m+1}	. . . .	x_n	
x	z_{m+1}	. . . .	z_n	z_0

where x is a variable which ought to have an integer value, but where its present value, z_0, is fractional. Our procedure then depends on whether the case is one of pure, or of mixed integer programming.

(a) Pure Integer Programming

Divide the z values into the largest integer contained in them, and a non-negative fraction, thus:

$$z_i = N_i + f_i \qquad (i = 0, m+1, \ldots, z_n)$$

Introduce a new non-negative, integer variable s by adding the row

$$s \quad -f_{m+1} \quad . \quad . \quad . \quad . \quad -f_n \quad -f_0$$

and continue by the Dual Simplex Method.

(b) Mixed Integer Programming

Assume that the variables are numbered in such a way that the

$z_{m+1}, \ldots, z_{m+k}$ are non-negative, and the
$z_{m+k+1}, \ldots, z_n$ are negative.

Divide z_0 (only) into the largest integer value contained in it, and a positive fraction, thus: $N_0 + f_0$. We shall need also the positive value $g_0 = f_0/(1 - f_0)$.

Introduce a new non-negative variable s by adding the row

$$s \quad -z_{m+1} \quad . \quad . \quad . \quad . \quad -z_{m+k} g_0 z_{m+k+1} \quad . \quad . \quad . \quad g_0 z_n \quad -f_0$$

and continue by the Dual Simplex Method.

Quadratic Programming

By this term we mean the problem of minimizing a convex quadratic function, subject to linear constraints, with non-negative variables. We describe here two methods, one due to E. M. L. Beale, and the other to G. B. Dantzig and C. van de Panne.

(a) Beale's Method

Let the quadratic function be written in terms of non-basic variables

$$C = \sum_{i=0}^{k} \sum_{j=0}^{k} a_{ij}x_i x_j + 2\sum_{i=0}^{k} \sum_{j=1}^{v} b_{ij}x_i u_j + \sum_{i=1}^{v} \sum_{j=1}^{v} c_{ij}u_i u_j$$

$a_{ij} = a_{ji}$, $c_{ij} = c_{ji}$, $x_0 = 1$, where all x_i must be non-negative, while the u_j may be positive, negative, or zero. It is convenient to write C as

$$\begin{array}{l} a_{00} + a_{01}x_1 + \ldots + a_{0k}x_k + b_{01}u_1 + \ldots + b_{0v}u_v \\ + [a_{10} + a_{11}x_1 + \ldots + a_{1k}x_k + b_{11}u_1 + \ldots + b_{1v}u_v]x_1 \\ \cdot \quad \cdot \quad \cdot \quad \cdot \quad \cdot \quad \cdot \quad \cdot \quad \cdot \\ + [a_{k0} + a_{k1}x_1 + \ldots + a_{kk}x_k + b_{k1}u_1 + \ldots + b_{kv}u_v]x_k \\ + [c_{11}u_1 + \ldots + c_{1v}u_v + b_{01} + b_{11}x_1 + \ldots + b_{k1}x_k]u_1 \\ \cdot \quad \cdot \quad \cdot \quad \cdot \quad \cdot \quad \cdot \quad \cdot \quad \cdot \\ + [c_{v1}u_1 + \ldots + c_{vv}u_v + b_{0v} + b_{1v}x_1 + \ldots + b_{kv}x_k]u_v \end{array}$$

Then the square brackets contain one half of the partial derivatives of C with regard to the variables outside the brackets.

Let $\partial C/\partial x_i = p_i$, say, and $\partial C/\partial u_j = q_j$.

If some $p_i < 0$, then we can increase x_i usefully, until either a constraint forces us to stop increasing it any further, or until $p_i = 0$, whichever happens earlier. We shall exchange x_i

in the former case for another x variable, or
in the latter case for a new variable, which we introduce as $u = pi$

If some $q_j \neq 0$, then we can change u_j in a useful direction, until either q_j or some x variable becomes zero. In the former case we exchange u_j for a new variable, which we introduce as

$$u = q_j$$

and in the latter case we exchange it for that x variable. The u variables will be distinguished by attaching consecutive subscripts to them.

Beale has shown in [4] that the routine terminates if we choose to make a u variable basic (positive or negative) whenever its partial derivative is not zero.

(b) The Dantzig–van de Panne Method

Let the quadratic form be written

$$C = \sum_{i=0}^{n} \sum_{j=0}^{n} c_{ij} x_i x_j \qquad x_0 = 1,\ c_{ij} = c_{ji}$$

and let the constraints be

$$\sum_{i=1}^{n} a_{ij} x_i - x_{n+j} = b_j \qquad (j = 1, \ldots, m)$$

$$x_i \geqslant 0,\ x_{n+j} \geqslant 0$$

Introduce y variables by

$$2 \sum_{j=0}^{n} c_{ij} x_j - \sum_{j=1}^{m} y_{n+j} a_{ij} - y_i = 0 \qquad (i = 1, \ldots, n)$$

$$y_i \geqslant 0,\ y_{n+j} \geqslant 0$$

We then express the "basic" variables $y_1, \ldots, y_n$ and $x_{n+1}, \ldots, x_{n+m}$ as well as the objective function in terms of the other variables in a "set-up tableau:"

	y_{n+1}	. .	y_{n+m}	x_1	. .	x_n	
y_1	a_{11}	. .	a_{1m}	$-2c_{11}$	. .	$-2c_{1n}$	c_{10}
.	.	.	.				
.							
y_n	a_{n1}	. .	a_{nm}	$-2c_{n1}$	. .	$-2c_{nn}$	c_{n0}
x_{n+1}	0	. .	0	$-a_{11}$	. .	$-a_{n1}$	$-b_1$
.	.	.	.				
.							
x_{n+m}	0	. .	0	$-a_{1m}$	. .	$-an_n$	$-b_m$
	$-b_1$	. .	$-b_m$	$-c_{10}$	. .	$-c_{n0}$	$2c_{00}$

This is a "Standard Tableau," i.e. of any pair (x_i, y_i) or (x_{n+j}, y_{n+j}) only one variable is basic, and the other is non-basic. During the further progress of the routine we might meet other Standard Tableaux, but also Non-standard ones, in which both variables of *one* such pair will be basic, and both of another pair non-basic.

The tableaux are transformed according to the Simplex rules, and the variables to be exchanged are determined by the following rules:

	Variable to become basic	Variable to become non-basic
Standard Tableaux	The x variable with the most negative y variable with the same subscript (the latter is in the basis)	Compare the ratios which are decisive in the Simplex Method, for all the x variables, and for the y variable of the new basic variable, provided the ratio is positive. Choose that with the smallest ratio.
Non-standard Tableaux	The y variable of the non-basic pair (x, y). There will be only one such pair.	Compare the ratios which are decisive in the Simplex Method, for all the x variables, and for the y variable of the basic pair (x, y), provided the ratio is positive. Choose that with the smallest ratio.

The procedure terminates when all y variables are non-negative. The x variables remain feasible throughout.

For detailed explanations and proofs see concerning

The Simplex Method and its variants: [12] and [35]
Integer Programming: [12], [22], and [23]
Quadratic Programming:
Beale's method [4] and [37]
Dantzig–van de Panne method [12] and [36].

Bibliography

[1] Allan, R. G. D., *Mathematical Economics* (London 1959) (2nd ed.).

[2] Balas, E., An additive algorithm for solving linear programs with zero-one variables, *Operations Research* **13** (1965) 517–46.

[3] Balas, E. and Ivanescu, P., On the generalized transportation problem, *Management Science* **11** (1964) 188–202.

[4] Beale, E. M. L., On quadratic programming, *Naval Research Logistics Quarterly* **6** (1959) 227–43.

[5] Bellman, R., *Dynamic Programming* (Princeton 1957).

[6] Bowman, E. H., Production scheduling by the transportation method of linear programming, *JORSA* **4** (1956) 100–103.

[7] Charnes, A. and Cooper, W. W., The stepping stone method of explaining linear programming calculations in transportation problems, *Management Science* **1** (1954) 49–89.

[8] Charnes, A. and Cooper, W. W., Management models and industrial applications of linear programming, *Management Science* **4** (1957) 38–91.

[9] Charnes, A. and Cooper, W. W., Programming with linear fractional functionals, *Naval Research Logistics Quarterly* **9** (1962) 181–86.

[10] Charnes, A. and Cooper, W. W., A network interpretation and a directed subdual algorithm for critical path scheduling, *Journal of Industrial Engineering* **13** (1962) 213–19.

[11] Charnes, A., Cooper, W. W. and Mellon, B., Blending aviation gasolines—a study in programming interdependent activities in an integrated oil company, *Econometrica* **20** (1952) 135–59.

[12] Dantzig, G. B., *Linear Programming and Extensions* (Princeton 1963).

[13] Derman, C. and Klein, M., Some remarks on finite horizon Markovian decision models. *Management Science* **13** (1965) 272–78.

[14] Eisemann, K., Simplified treatment of degeneracy in transportation problems, *Quarterly of Applied Mathematics* **14** (1957) 399–403.

[15] Eisemann, K., The trim problem, *Management Science* **3** (1957) 279–84.

[16] Eisemann, K., The generalized stepping stone method for the machine loading model, *Management Science* **11** (1964) 154–76.

[17] Ferguson, A. R. and Dantzig, G. B., The problem of routing aircraft, *Aeronautical Engineering Review* **14** (1955) 51–55.

[18] Ford, L. R. jr. and Fulkerson, D. R., Solving the transportation problem, *Management Science* **3** (1956) 24–32.

[19] Ford, L. R. jr. and Fulkerson, D. R., *Flow in Networks* (Princeton 1962).

[20] Gilmore, P. C. and Gomory, R. E., A linear programming approach to the cutting stock problem, Part II, *Operations Research* **11** (1963) 863–88.

[21] Girault, M., Stratégie de lancement, *Revue de statistique appliquée* **13** (1965) 61–68.

[22] Gomory, R. E., An algorithm for integer solutions to linear programs, *Recent Advances in Mathematical Programming*, Edit. R. L. Graves and P. Wolfe, 269–302 (New York 1962).

[23] Gomory, R. E., An algorithm for the mixed integer problem, *RM*–2597 (Rand Corporation Report 1960).

[24] Hall, P., On representations of subsets, *Journal of the London Mathematical Society* **10** (1935) 26–30.

[25] Hitchcock, F. L., The distribution of a product from several sources to numerous localities, *Journal for Mathematics and Physics* **20** (1941) 224–230.

[26] Howard, R., *Dynamic Programming and Markov Processes* (New York 1960).

[27] Kelley, J. E. jr., Critical path planning and scheduling: mathematical basis, *Operations Research* **9** (1961) 296–320.

[28] Kuhn, H. W. and Tucker, A. W., Preface to Annals of Mathematics Studies 38, *Linear Inequalities and Related Systems* (Princeton 1956).

[29] Laderman, J. and Weiss, L., The inventory problem, *Journal of the American Statistical Association* **48** (1953) 717–32.

[30] Lemke, C. E., The dual method of solving linear programming problems. *Naval Research Logistics Quarterly* **1** (1954) 36–47.

[31] Lourie, Janice R., Topology and computation of the generalized transportation problem, *Management Science* **11** (1964) 177–87.

[32] Orden, A., The transhipment problem, *Management Science* **2** (1956) 276–85.

[33] Stanley, E. D., Honig, D. P. and Gainen, L., Linear programming in bid evaluation, *Naval Research Logistics Quarterly* **1** (1954) 48–52.

[34] Stigler, G. J., The cost of subsistence, *Journal of Farm Economics* **27** (1945) 303–14.

[35] Vajda, S., *Mathematical Programming* (Reading (Mass.), 1961).

[36] van de Panne, C. and Whinston, A., Simplicial methods for quadratic programming, *Naval Research Logistics Quarterly* **11** (1964) 273–302.

[37] van de Panne, C. and Whinston, A., A comparison of two methods for quadratic programming, *Operations Research* **14** (1966) 422–41. With a Note by E. M. L. Beale, 442–43.

[38] White, D. J., Dynamic programming and systems of uncertain duration, *Management Science* **12** (1965) 37–67.

[39] Wolfe, P. and Dantzig, G. B., Linear programming in a Markov chain, *Operations Research* **10** (1962) 702–10.

[40] Jewell, W. S., Markov–Renewal Programming, I. *Operations Research* **11** (1963) 938–48, II. *ibid* 949–71.

[41] Ford, L. R. jr. and Fulkerson, D. R., A suggested computation for maximal multi-commodity flows, *Management Science* **5** (1958) 97–101.